Hands-On Natural Language Processing with PyTorch 1.x

Build smart, AI-driven linguistic applications using deep learning and NLP techniques

Thomas Dop

BIRMINGHAM—MUMBAI

Hands-On Natural Language Processing with PyTorch 1.x

Commissioning Editor: Amey Varangaonkar

Acquisition Editor: Devika Battike

Senior Editor: David Sugarman

Content Development Editor: Joseph Sunil

Technical Editor: Manikandan Kurup

Copy Editor: Safis Editing

Project Coordinator: Aishwarya Mohan

Proofreader: Safis Editing

Indexer: Rekha Nair

Production Designer: Jyoti Chauhan

First published: July 2020

Production reference: 1080720

Published by Packt Publishing Ltd.
Livery Place
35 Livery Street
Birmingham
B3 2PB, UK.

ISBN 978-1-78980-274-0

www.packt.com

For Mhairi and Dr. F.R. Allen

–Thomas Dop

Contributors

About the author

Thomas Dop is a data scientist at MagicLab, a company that creates leading dating apps, including Bumble and Badoo. He works on a variety of areas within data science, including NLP, deep learning, computer vision, and predictive modeling. He holds an MSc in data science from the University of Amsterdam.

About the reviewers

Nilan Saha is pursuing a Master's degree in Data Science with a specialization in Computational Linguistics from the University of British Columbia, Canada. He has worked as an NLP contractor for multiple startups in the past, and has also got brief experience in research, which has resulted in a few publications. He is also a Kaggle Kernels and Discussion Expert.

Chintan Gajjar is a senior consultant in KNOWARTH Technologies. He has also contributed to books such as Hadoop Backup and Recovery Solutions, MySQL 8 for Big Data, MySQL 8 Administrator's Guide, and Hands-on Natural Language Processing with Python. He has a Master's degree in computer applications from Ganpat University, India.

I would like to thank the author, co-reviewer, and the wonderful team at Packt Publishing for all efforts and my office colleagues, Darshan Kansara and Kathan Thakkar, for supporting me throughout the reviewing of this book. They both are technology enthusiasts and have a great understanding of AI/ML, CI-CD, and are great mentors.

Packt is searching for authors like you

If you're interested in becoming an author for Packt, please visit `authors.packtpub.com` and apply today. We have worked with thousands of developers and tech professionals, just like you, to help them share their insight with the global tech community. You can make a general application, apply for a specific hot topic that we are recruiting an author for, or submit your own idea.

Table of Contents

Section 2: Fundamentals of Natural Language Processing

In this section, you will learn about the fundamentals of building a **Natural Language Processing** (**NLP**) application. You will also learn how to use various NLP techniques, such as word embeddings, CBOW, and tokenization in PyTorch in this section.

3

NLP and Text Embeddings

4

Text Preprocessing, Stemming, and Lemmatization

Section 3:
Real-World NLP Applications Using PyTorch 1.x

5
Recurrent Neural Networks and Sentiment Analysis

6
Convolutional Neural Networks for Text Classification

7

Text Translation Using Sequence-to-Sequence Neural Networks

8

Building a Chatbot Using Attention-Based Neural Networks

9
The Road Ahead

Other Books You May Enjoy

Index

Preface

In the internet age, where an increasing volume of text data is being generated daily from social media and other platforms, being able to make sense of that data is a crucial skill. This book will help you build deep learning models for **Natural Language Processing (NLP)** tasks that will help you extract valuable insights from text.

We will start by understanding how to install PyTorch and using CUDA to accelerate the processing speed. You'll then explore how the NLP architecture works through practical examples. Later chapters will guide you through important principles, such as word embeddings, CBOW, and tokenization in PyTorch. You'll then learn some techniques for processing textual data and how deep learning can be used for NLP tasks. Next, we will demonstrate how to implement deep learning and neural network architectures to build models that allow you to classify and translate text and perform sentiment analysis. Finally, you will learn how to build advanced NLP models, such as conversational chatbots.

By the end of this book, you'll understand how different NLP problems can be solved using deep learning with PyTorch, as well as how to build models to solve them.

Who this book is for

This PyTorch book is for NLP developers, machine learning and deep learning developers, or anyone working toward building intelligent language applications using both traditional NLP approaches and deep learning architectures. If you're looking to adopt modern NLP techniques and models for your development projects, then this book is for you. Working knowledge of Python programming and basic working knowledge of NLP tasks are a must.

What this book covers

Chapter 1, Fundamentals of Machine Learning and Deep Learning, provides an overview of the fundamental aspects of machine learning and neural networks.

Chapter 2, Getting Started with PyTorch 1.x for NLP, shows you how to download, install, and start PyTorch. We will also run through some of the basic functionality of the package.

Chapter 3, NLP and Text Embeddings, shows you how to create text embeddings for NLP and use them in basic language models.

Chapter 4, Text Preprocessing, Stemming, and Lemmatization, shows you how to preprocess textual data for use in NLP deep learning models.

Chapter 5, Recurrent Neural Networks and Sentiment Analysis, runs through the fundamentals of recurrent neural networks and shows you how to use them to build a sentiment analysis model from scratch.

Chapter 6, Convolutional Neural Networks for Text Classification, runs through the fundamentals of convolutional neural networks and shows you how you can use them to build a working model for classifying text.

Chapter 7, Text Translation Using Sequence-to-Sequence Neural Networks, introduces the concept of sequence-to-sequence models for deep learning and runs through how to use them to construct a model to translate text into another language.

Chapter 8, Building a Chatbot Using Attention-Based Neural Networks, covers the concept of attention for use within sequence-to-sequence deep learning models and also shows you how they can be used to build a fully working chatbot from scratch.

Chapter 9, The Road Ahead, covers some of the state-of-the-art models currently used within NLP deep learning and looks at some of the challenges and problems facing the field of NLP going forward.

To get the most out of this book

You will need a version of Python installed on your computer. All code examples have been tested using version 3.7. You will also need a working PyTorch environment for the deep learning components of this book. All deep learning models were constructed using version 1.4; however, the majority of the code should work with later versions.

Software/Hardware covered in the book	OS Requirements
Python 3.7	Windows/Linux/macOS
PyTorch 1.4	Windows/Linux/macOS

There are several Python libraries used within the code throughout this book; however, these will be covered in the relevant chapters.

If you are using the digital version of this book, we advise you to type the code yourself or access the code via the GitHub repository (link available in the next section). Doing so will help you avoid any potential errors related to the copying and pasting of code.

Download the example code files

You can download the example code files for this book from your account at www.packt.com. If you purchased this book elsewhere, you can visit www.packtpub.com/support and register to have the files emailed directly to you.

You can download the code files by following these steps:

1. Log in or register at www.packt.com.
2. Select the **Support** tab.
3. Click on **Code Downloads**.
4. Enter the name of the book in the **Search** box and follow the onscreen instructions.

Once the file is downloaded, please make sure that you unzip or extract the folder using the latest version of:

- WinRAR/7-Zip for Windows
- Zipeg/iZip/UnRarX for Mac
- 7-Zip/PeaZip for Linux

The code bundle for the book is also hosted on GitHub at `https://github.com/ PacktPublishing/Hands-On-Natural-Language-Processing-with- PyTorch-1.x`. In case there's an update to the code, it will be updated on the existing GitHub repository.

We also have other code bundles from our rich catalog of books and videos available at `https://github.com/PacktPublishing/`. Check them out!

Download the color images

We also provide a PDF file that has color images of the screenshots/diagrams used in this book. You can download it here: `https://static.packt-cdn.com/ downloads/9781789802740_ColorImages.pdf`.

Conventions used

There are a number of text conventions used throughout this book.

`Code in text`: Indicates code words in text, database table names, folder names, filenames, file extensions, pathnames, dummy URLs, user input, and Twitter handles. Here is an example: "Mount the downloaded `WebStorm-10*.dmg` disk image file as another disk in your system."

A block of code is set as follows:

```
import torch
```

When we wish to draw your attention to a particular part of a code block, the relevant lines or items are set in bold:

```
word_1 = 'cat'
word_2 = 'dog'
word_3 = 'bird'
```

Any command-line input or output is written as follows:

```
$ mkdir flaskAPI
$ cd flaskAPI
```

Bold: Indicates a new term, an important word, or words that you see onscreen. For example, words in menus or dialog boxes appear in the text like this. Here is an example: "Select **System info** from the **Administration** panel."

> **Tips or important notes**
> Appear like this.

Get in touch

Feedback from our readers is always welcome.

General feedback: If you have questions about any aspect of this book, mention the book title in the subject of your message and email us at `customercare@packtpub.com`.

Errata: Although we have taken every care to ensure the accuracy of our content, mistakes do happen. If you have found a mistake in this book, we would be grateful if you would report this to us. Please visit www.packtpub.com/support/errata, selecting your book, clicking on the Errata Submission Form link, and entering the details.

Piracy: If you come across any illegal copies of our works in any form on the Internet, we would be grateful if you would provide us with the location address or website name. Please contact us at copyright@packt.com with a link to the material.

If you are interested in becoming an author: If there is a topic that you have expertise in and you are interested in either writing or contributing to a book, please visit authors. packtpub.com.

Reviews

Please leave a review. Once you have read and used this book, why not leave a review on the site that you purchased it from? Potential readers can then see and use your unbiased opinion to make purchase decisions, we at Packt can understand what you think about our products, and our authors can see your feedback on their book. Thank you!

For more information about Packt, please visit packt.com.

Section 1: Essentials of PyTorch 1.x for NLP

In this section, you will learn about the basic concepts of PyTorch 1.x in the context of **Natural Language Processing (NLP)**. You will also learn how to install PyTorch 1.x on your machine, as well as how to use CUDA to accelerate the processing speed.

This section contains the following chapters:

- *Chapter 1, Fundamentals of Machine Learning and Deep Learning*
- *Chapter 2, Getting Started with PyTorch 1.x for NLP*

1
Fundamentals of Machine Learning and Deep Learning

Our world is rich with natural language data. Over the past several decades, the way we communicate with one another has shifted to the digital realm and, as such, this data can be used to build models that can improve our online experience. From returning relevant results within a search engine, to autocompleting the next word you type in an email, the benefits of being able to extract insights from natural language is clear to see.

While the way we, as humans, understand language differs notably from the way a model or *artificial intelligence* may understand it, by shedding light on machine learning and what it is used for, we can begin to understand just how these deep learning models *understand* language and what fundamentally happens when a model learns from data.

Throughout this book, we will explore this application of artificial intelligence and deep learning to natural language. Through the use of PyTorch, we will learn, step by step, how to build models that allow us to perform sentiment analysis, text classification, and sequence translation, which will lead to us building a basic chatbot. By covering the theory behind each of these models, as well as demonstrating how to implement them practically, we will demystify the field of **natural language processing** (**NLP**) and provide you with enough background for you to start building your own models.

In our first chapter, we will explore some of the basic concepts of machine learning. We will then take this a step further by examining the fundamentals of deep learning, neural networks, and some of the advantages that deep learning methods afford over basic machine learning techniques. Finally, we will take a more detailed look at deep learning, specifically with regard to NLP-specific tasks and how we can use deep learning models to gain insights from natural language. Specifically, we'll cover the following topics:

- Overview of machine learning
- Introduction to neural networks
- NLP for machine learning

Overview of machine learning

Fundamentally, machine learning is the algorithmic process used to identify patterns and extract trends from data. By training specific machine learning algorithms on data, a machine learning model may learn insights that aren't immediately obvious to the human eye. A medical imaging model may learn to detect cancer from images of the human body, while a sentiment analysis model may learn that a book review containing the words *good*, *excellent*, and *entertaining* is more likely to be a positive review than one containing the words *bad*, *terrible*, and *boring*.

Broadly speaking, machine learning algorithms fall into two main categories: supervised learning and unsupervised learning.

Supervised learning

Supervised learning covers any task where we wish to use an input to predict an output. Let's say we wish to train a model to predict house prices. We know that larger houses tend to sell for more money, but we don't know the exact relationship between price and size. A machine learning model can learn this relationship by looking at the data:

Size (m sq)	Price ($)
100	320,000
150	440,000
130	400,000
120	350,000
125	?

Figure 1.1 – Table showing housing data

Here, we have been given the sizes of four houses that recently sold, as well as the prices they sold for. Given the data on these four houses, can we use this information to make a prediction about a new house on the market? A simple machine learning model known as a **regression** can estimate the relationship between these two factors:

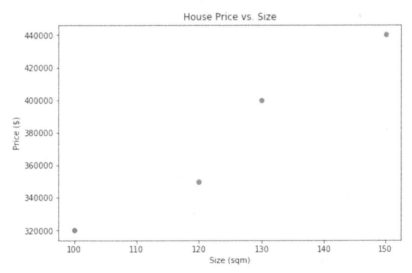

Figure 1.2 – Output of the housing data

Given this historic data, we can use this data to estimate a relationship between **size** (X) and **price** (Y). Now that we have an estimation of the relationship between size and price, if we are given a new house where we just know its size, we can use this to predict its price using the learned function:

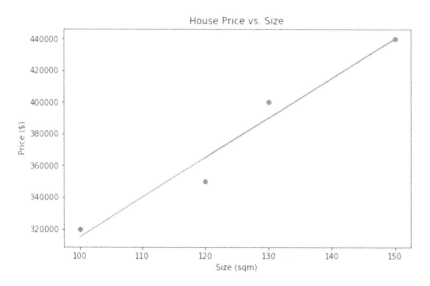

Figure 1.3 – Predicting house prices

Therefore, all supervised learning tasks aim to learn some function of the model inputs to predict an output, given many examples of how input relates to output:

Given many (X, y), learn:

$F(X) = y$

The input to your number can consist of any number of features. Our simple house price model consisted of just a single feature (**size**), but we may wish to add more features to give an even better prediction (for example, number of bedrooms, size of the garden, and so on). So, more specifically, our supervised model learns a function in order to map a number of inputs to an output. This is given by the following equation:

Given many ([X0, X1, X2,...,Xn], y), learn:

$f(X0, X1, X2,...,Xn) = y$

In the preceding example, the function that we learn is as follows:

$$y = \theta_0 + \theta_1 x$$

Here, θ_0 is the x axis intercept and θ_1 is the slope of the line.

Models can consist of millions, even billions, of input features (though you may find that you run into hardware limitations when the feature space becomes too large). The types of inputs to a model may vary as well, with models being able to learn from images:

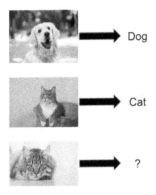

Figure 1.4 – Model training

As we shall explore in more detail later, they can also learn from text:

I loved this film -> Positive

This movie was terrible -> Negative

The best film I saw this year -> ?

Unsupervised learning

Unsupervised learning differs from supervised learning in that unsupervised learning doesn't use pairs of inputs and outputs (*X, y*) to learn. Instead, we only provide input data and the model will learn something about the structure or representation of the input data. One of the most common methods of unsupervised learning is **clustering**.

For example, we take a dataset of readings of temperatures and rainfall measures from a set of four different countries but have no labels about where these readings were taken. We can use a clustering algorithm to identify the distinct clusters (countries) that exist within the data:

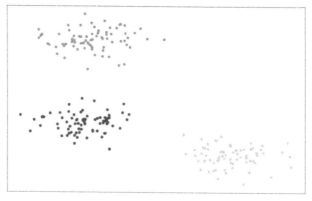

Figure 1.5 – Output of the clustering algorithm

Clustering also has uses within the realm of NLP. If we are given a dataset of emails and want to determine how many different languages are being spoken within these emails, a form of clustering could help us identify this. If English words appear frequently with other English words within the same email and Spanish words appear frequently with other Spanish words, we would use clustering to determine how many distinct clusters of words our dataset has and, thus, the number of languages.

How do models learn?

In order for models to learn, we need some way of evaluating how our model is performing. To do this, we use a concept called loss. **Loss** is a measure of how close our model predictions are from their true values. For a given house in our dataset, one measure of loss could be the difference between the true price (y) and the price predicted by our model ($\bar{y}$). We could assess the total loss within our system by taking an average of this loss across all houses in the dataset. However, the positive loss could theoretically cancel out negative loss, so a more common measure of loss is the **mean squared error**:

$$MSE = \frac{1}{n}\sum (y - \bar{y})^2$$

While other models may use different loss functions, regressions generally use mean squared error. Now, we can calculate a measure of loss across our entire dataset, but we still need a way of algorithmically arriving at the lowest possible loss. This process is known as **gradient descent**.

Gradient descent

Here, we have plotted our loss function as it relates to a single learned parameter within our house price model, θ_1. We note that when θ_1 is set too high, the MSE loss is high, and when θ_1 is set too low, the MSE loss is also high. The *sweet spot*, or the point where the loss is minimized, lies somewhere in the middle. To calculate this algorithmically, we use gradient descent. We will see this in more detail when we begin to train our own neural networks:

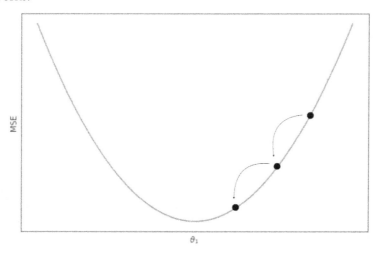

Figure 1.6 – Gradient descent

We first initialize θ_1 with a random value. To reach the point where our loss is minimized, we need to move further *downhill* from our loss function, toward the middle of the valley. To do this, we first need to know which direction to move in. At our initial point, we use basic calculus to calculate the initial gradient of the slope:

$$Gradient = \frac{\partial MSE}{\partial \theta_1}$$

In our preceding example, the gradient at the initial point is positive. This tells us that our value of θ_1 is larger than the optimal value, so we update our value of θ_1 so that it's lower than our previous value. We gradually iterate this process until θ_1 moves closer and closer to the value where MSE is minimized. This happens at the point where the gradient equals zero.

Overfitting and underfitting

Consider the following scenario, where a basic linear model is poorly fitted to our data. We can see that our model, denoted by the equation $y = \theta_0 + \theta_1 X$, does not appear to be a good predictor:

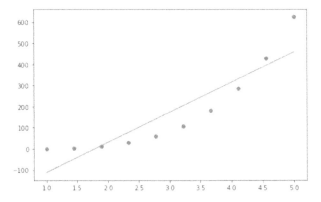

Figure 1.7 – Example of underfitting and overfitting

When our model does not fit the data well because of a lack of features, lack of data, or model underspecification, we call this **underfitting**. We note the increasing gradient of our data and suspect that a model, if using a polynomial, may be a better fit; for example, $y = \theta_0 + \theta_1 X + \theta_2 X^2$. We will see later that due to the complex architecture of neural networks, underfitting is rarely an issue:

Consider the following example. Here, we're fitting a function using our house price model to not only the size of the house (*X*), but the second and third order polynomials too *(X2, X3)*. Here, we can see that our new model fits our data points perfectly. However, this does not necessarily result in a good model:

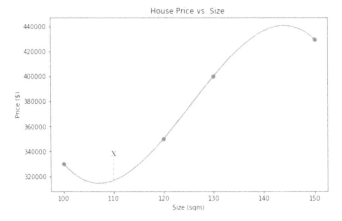

Figure 1.8 – Sample output of overfitting

We now have a house of size **110 sq m** to predict the price of. Using our intuition, as this house is larger than the **100 sq m** house, we would expect this house to be more expensive at around **$340,000**. Using our fitted polynomial model, we can see that the predicted price is actually lower than the smaller house at around **$320,000**. Our model fits the data we have trained it on well, but it does not generalize well to a new, unseen datapoint. This is known as **overfitting**. Because of overfitting, it is important not to evaluate a model's performance on the data it was trained on, so we need to generate a separate set of data to evaluate our data on.

Train versus test

Normally, when training models, we separate our data into two parts: a training set of data and a smaller test set of data. We train the model using the training set of data and evaluate it on the test set of data. This is done in order to measure the model's performance on an unseen set of data. As mentioned previously, for a model to be a good predictor, it must generalize well to a new set of data that the model hasn't seen before, and this is precisely what evaluating on a testing set of data measures.

Evaluating models

While we seek to minimize loss in our models, this alone does not give us much information about how good our model is at actually making predictions. Consider an anti-spam model that predicts whether a received email is spam or not and automatically sends spam emails to a junk folder. One simple measure of evaluating performance is **accuracy**:

$$Accuracy = \frac{number\ of\ correct\ predictions}{total\ predictions}$$

To calculate accuracy, we simply take the number of emails that were predicted correctly as spam/non-spam and divide this by the total number of predictions we made. If we correctly predicted 990 emails out of 1,000, we would have an accuracy of 99%. However, a high accuracy does not necessarily mean our model is good:

	Predicted Spam	Predicted Not Spam
Is Spam	0	10
Is Not Spam	0	990

Figure 1.9 – Table showing data predicted as spam/non-spam

Here, we can see that although our model predicted 990 emails as not spam correctly (known as true negatives), it also predicted 10 emails that were spam as not spam (known as false negatives). Our model just assumes that all emails are not spam, which is not a good anti-spam filter at all! Instead of just using accuracy, we should also evaluate our model using **precision and recall**. In this scenario, the fact that our model would have a recall of zero (meaning no positive results were returned) would be an immediate red flag:

$$Precision = \frac{True\ Positives}{True\ Positives + False\ Positives}$$

$$Recall = \frac{True\ Positives}{True\ Positives + False\ Negatives}$$

Neural networks

In our previous examples, we have discussed mainly regressions in the form $y = \theta_0 + \theta_1 X$. We have touched on using polynomials to fit more complex equations such as $y = \theta_0 + \theta_1 X + \theta_2 X$. However, as we add more features to our model, when to use a transformation of the original feature becomes a case of trial and error. Using **neural networks**, we are able to fit a much more complex function, $y = f(X)$, to our data, without the need to engineer or transform our existing features.

Structure of neural networks

When we were learning the optimal value of θ_1, which minimized loss in our regressions, this is effectively the same as a **one-layer neural network**:

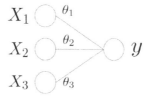

Figure 1.10 – One-layer neural network

Here, we take each of our features, X_i, as an input, illustrated here by a **node**. We wish to learn the parameters, θ_i, which are represented as **connections** in this diagram. Our final sum of all the products between X_i and θ_i gives us our final prediction, y:

$$\sum X_i \theta_i = X_1 \theta_1 + X_2 \theta_2 + X_3 \theta_3 = y$$

A neural network simply builds upon this initial concept, adding extra layers to the calculation, thus increasing the complexity and the parameters learned, giving us something like this:

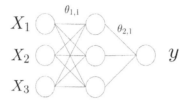

Figure 1.11 – Fully connected network

Every input node is connected to every node in another layer. This is known as a **fully connected layer**. The output from the fully connected layer is then multiplied by its own additional weights in order to predict y. Therefore, our predictions are no longer just a function of $X_i \theta_i$ but now include multiple learned weights against each parameter. Feature X_1 is no longer affected not just by θ_1. Now, it is also affected by the $\theta_{1,1}, \theta_{2,1}, \theta_{2,2}$ and $\theta_{2,3}$. parameters.

Since each node within the fully connected layer takes all values of X as input, the neural network is able to learn interaction features between the input features. Multiple fully connected layers can be chained together to learn even more complex features. In this book, we will see that all of the neural networks we build will use this concept; chaining together multiple layers of different varieties in order to construct even more complex models. However, there is one additional key element to cover before we can fully understand neural networks: activation functions.

Activation functions

While chaining various weights together allows us to learn more complex parameters, ultimately, our final prediction will still be a combination of the linear products of weights and features. If we wish our neural networks to learn a truly complex, non-linear function, then we must introduce an element of nonlinearity into our model. This is done through the use of **activation functions**:

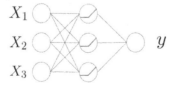

Figure 1.12 – Activation functions in neural networks

We apply an activation function to each node within our fully connected layer. What this means is that each node in the fully connected layer takes a sum of features and weights as input, applies a nonlinear function to the resulting value, and outputs the transformed result. While there are many different activation functions, the most frequently used in recent times is **ReLU**, or the **Rectified Linear Unit**:

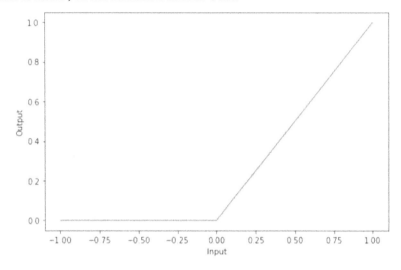

Figure 1.13 – Representation of ReLU output

ReLU is a very simple non-linear function that returns $y = 0$ when $y \leq 0$ and $y = X$ when $X > 0$. After introducing these activation functions to our model, our final learned function becomes nonlinear, meaning we can create more models than we would have been able to using a combination of conventional regression and feature engineering alone.

How do neural networks learn?

The act of learning from our data using neural networks is slightly more complicated than when we learned using basic regressions. While we still use gradient descent as before, the actual loss function we need to differentiate becomes significantly more complex. In a one-layered neural network with no activation functions, we can easily calculate the derivative of the loss function as it is easy to see how the loss function changes as we vary each parameter. However, in a multi-layered neural network with activation functions, this is more complex.

We must first perform a **forward-pass**, which is where, using the model's current state, we compute the predicted value of y and evaluate this against the true value of y in order to obtain a measure of loss. Using this loss, we move backward through the network, calculating the gradient at each parameter within the network. This allows us to know which direction to update our parameter in so that we can move closer toward the point where loss is minimized. This is known as **backpropagation**. We can calculate the derivative of the loss function with respect to each parameter using the **chain rule**:

$$\frac{\partial L}{\partial \theta_{i,j}} = \frac{\partial L}{\partial o_i} \frac{\partial o_j}{\partial \theta_{i,j}}$$

Here, o_j is the output at each given node within the network. So, to summarize, the four main steps we take when performing gradient descent on neural networks are as follows:

1. Perform a forward pass using your data, calculating the total loss of the network.

2. Using backpropagation, calculate the gradients of each parameter with respect to loss at each node in the network.

3. Update the values of these parameters, moving toward the direction where loss is minimized.

4. Repeat until convergence.

Overfitting in neural networks

We saw that, in the case of our regressions, it was possible to add so many features that it was possible to overfit the network. This gets to a point where the model fits the training data perfectly but does not generalize well to an unseen test set of data. This is a common problem in neural networks as the increased complexity of the models means that it is often possible to fit a function to the training set of data that doesn't necessarily generalize. The following is a plot of the total loss on the training and test sets of data after each forward and backward pass of the dataset (known as an epoch):

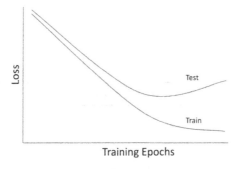

Figure 1.14 – Test and training epochs

Here, we can see that as we continue to train the network, the training loss gets smaller over time as we move closer to the point where the total loss is minimized. While this generalizes well to the test set of data up to a point, after a while, the total loss on the test set of data begins to increase as our function overfits to the data in the training set. One solution to this is **early stopping**. Because we want our model to make good predictions on data it hasn't seen before, we can stop training our model at the point where test loss is minimized. A fully trained NLP model may be able to easily classify sentences it has seen before, but the measure of a model that has truly learned something is its ability to make predictions on unseen data.

NLP for machine learning

Unlike humans, computers do not understand text – at least not in the same way that we do. In order to create machine learning models that are able to learn from data, we must first learn to represent natural language in a way that computers are able to process.

When we discussed machine learning fundamentals, you may have noticed that loss functions all deal with numerical data so as to be able to minimize loss. Because of this, we wish to represent our text in a numerical format that can form the basis of our input into a neural network. Here, we will cover a couple of basic ways of numerically representing our data.

Bag-of-words

The first and most simple way of representing text is by using a **bag-of-words** representation. This method simply counts the words in a given sentence or document and counts all the words. These counts are then transformed into a vector where each element of the vector is the count of the times each word in the **corpus** appears within the sentence. The corpus is simply all the words that appear across all the sentences/documents being analyzed. Take the following two sentences:

The cat sat on the mat

The dog sat on the cat

We can represent each of these sentences as a count of words:

	the	cat	dog	sat	on	mat
Sentence 1	2	1	0	1	1	1
Sentence 2	2	1	1	1	1	0

Figure 1.15 – Table of word counts

Then, we can transform these into individual vectors:

The cat sat on the mat -> [2,1,0,1,1,1]

The dog sat on the cat -> [2,1,1,1,1,0]

This numeric representation could then be used as the input features to a machine learning model where the feature vector is $[X_0, X_1, ..., X_n]$.

Sequential representation

We will see later in this book that more complex neural network models, including RNNs and LSTMs, do not just take a single vector as input, but can take a whole sequence of vectors in the form of a matrix. Because of this, in order to better capture the order of words and thus the meaning of any sentence, we are able to represent this in the form of a sequence of one-hot encoded vectors:

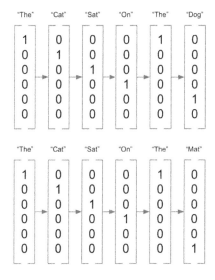

Figure 1.16 – One-hot encoded vectors

Summary

In this chapter, we introduced the fundamentals of machine learning and neural networks, as well as a brief overview of transforming text for use within these models. In the next chapter, we will provide a brief overview of PyTorch and how it can be used to construct some of these models.

2
Getting Started with PyTorch 1.x for NLP

PyTorch is a Python-based machine learning library. It consists of two main features: its ability to efficiently perform tensor operations with hardware acceleration (using GPUs) and its ability to build deep neural networks. PyTorch also uses dynamic computational graphs instead of static ones, which sets it apart from similar libraries such as TensorFlow. By demonstrating how language can be represented using tensors and how neural networks can be used to learn from NLP, we will show that both these features are particularly useful for natural language processing.

In this chapter, we will show you how to get PyTorch up and running on your computer, as well as demonstrate some of its key functionalities. We will then compare PyTorch to some other deep learning frameworks, before exploring some of the NLP functionality of PyTorch, such as its ability to perform tensor operations, and finally demonstrate how to build a simple neural network. In summary, this chapter will cover the following topics:

- Installing PyTorch
- Comparing PyTorch to other deep learning frameworks
- NLP functionality of PyTorch

Technical requirements

For this chapter, Python needs to be installed. It is recommended to use the latest version of Python (3.6 or higher). It is also recommended to use the Anaconda package manager to install PyTorch. A CUDA-compatible GPU is required to run tensor operations on a GPU. All the code for this chapter can be found at `https://github.com/PacktPublishing/Hands-On-Natural-Language-Processing-with-PyTorch-1.x`.

Installing and using PyTorch 1.x

Like most Python packages, PyTorch is very simple to install. There are two main ways of doing so. The first is to simply install it using `pip` in the command line. Simply type the following command:

```
pip install torch torchvision
```

While this installation method is quick, it is recommended to install using Anaconda instead, as this includes all the required dependencies and binaries for PyTorch to run. Furthermore, Anaconda will be required later to enable training models on a GPU using CUDA. PyTorch can be installed through Anaconda by entering the following in the command line:

```
conda install torch torchvision -c pytorch
```

To check that PyTorch is working correctly, we can open a Jupyter Notebook and run a few simple commands:

1. To define a Tensor in PyTorch, we can do the following:

    ```
    import torch
    x = torch.tensor([1.,2.])
    print(x)
    ```

    This results in the following output:

    ```
    tensor([1., 2.])
    ```

    Figure 2.1 – Tensor output

This shows that tensors within PyTorch are saved as their own data type (not dissimilar to how arrays are saved within NumPy).

2. We can perform basic operations such as multiplication using standard Python operators:

```
x = torch.tensor([1., 2.])
y = torch.tensor([3., 4.])
print(x * y)
```

This results in the following output:

```
tensor([3., 8.])
```

Figure 2.2 – Tensor multiplication output

3. We can also select individual elements from a tensor, as follows:

```
x = torch.tensor([[1., 2.],[5., 3.],[0., 4.]])

print(x[0][1])
```

This results in the following output:

```
tensor(2.)
```

Figure 2.3 – Tensor selection output

However, note that unlike a NumPy array, selecting an individual element from a tensor object returns another tensor. In order to return an individual value from a tensor, you can use the .item() function:

```
print(x[0][1].item())
```

This results in the following output:

```
2.0
```

Figure 2.4 – Output of the .item() function

Tensors

Before we continue, it is important that you are fully aware of the properties of a tensor. Tensors have a property known as an **order**, which essentially determines the dimensionality of a tensor. An order one tensor is a tensor with a single dimension, which is equivalent to a vector or list of numbers. An order 2 tensor is a tensor with two dimensions, equivalent to a matrix, whereas a tensor of order 3 consists of three dimensions. There is no limit to the maximum order a tensor can have within PyTorch:

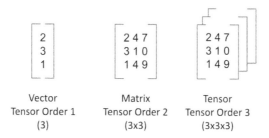

Figure 2.5 – Tensor matrix

You can check the size of any tensor by typing the following:

```
x.shape
```

This results in the following output:

```
torch.Size([3, 2])
```

Figure 2.6 – Tensor shape output

This shows that this is a 3x2 tensor (order 2).

Enabling PyTorch acceleration using CUDA

One of the main benefits of PyTorch is its ability to enable acceleration through the use of a **graphics processing unit** (**GPU**). Deep learning is a computational task that is easily parallelizable, meaning that the calculations can be broken down into smaller tasks and calculated across many smaller processors. This means that instead of needing to execute the task on a single CPU, it is more efficient to perform the calculation on a GPU.

GPUs were originally created to efficiently render graphics, but since deep learning has grown in popularity, GPUs have been frequently used for their ability to perform multiple calculations simultaneously. While a traditional CPU may consist of around four or eight cores, a GPU consists of hundreds of smaller cores. Because calculations can be executed across all these cores simultaneously, GPUs can rapidly reduce the time taken to perform deep learning tasks.

Consider a single pass within a neural network. We may take a small batch of data, pass it through our network to obtain our loss, and then backpropagate, adjusting our parameters according to the gradients. If we have many batches of data to do this over, on a traditional CPU, we must wait until batch 1 has completed before we can compute this for batch 2:

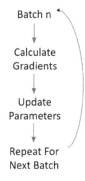

Figure 2.7 – One pass in a neural network

However, on a GPU, we can perform all these steps simultaneously, meaning there is no requirement for batch 1 to finish before batch 2 can be started. We can calculate the parameter updates for all batches simultaneously and then perform all the parameter updates in one go (as the results are independent of one another). The parallel approach can vastly speed up the machine learning process:

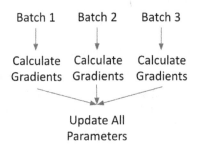

Figure 2.8 – Parallel approach to perform passes

Compute Unified Device Architecture (**CUDA**) is the technology specific to Nvidia GPUs that enables hardware acceleration on PyTorch. In order to enable CUDA, we must first make sure the graphics card on our system is CUDA-compatible. A list of CUDA-compatible GPUs can be found here: `https://developer.nvidia.com/cuda-gpus`. If you have a CUDA-compatible GPU, then CUDA can be installed from this link: `https://developer.nvidia.com/cuda-downloads`. We will activate it using the following steps:

1. Firstly, in order to actually enable CUDA support on PyTorch, you will have to build PyTorch from source. Details about how this can be done can be found here: `https://github.com/pytorch/pytorch#from-source`.

2. Then, to actually CUDA within our PyTorch code, we must type the following into our Python code:

    ```
    cuda = torch.device('cuda')
    ```

    This sets our default CUDA device's name to `'cuda'`.

3. We can then execute operations on this device by manually specifying the device argument in any tensor operations:

    ```
    x = torch.tensor([5., 3.], device=cuda)
    ```

    Alternatively, we can do this by calling the `cuda` method:

    ```
    y = torch.tensor([4., 2.]).cuda()
    ```

4. We can then run a simple operation to ensure this is working correctly:

    ```
    x*y
    ```

    This results in the following output:

    ```
    tensor([20.,   6.])
    ```

Figure 2.9 – Tensor multiplication output using CUDA

The changes in speed will not be noticeable at this stage as we are just creating a tensor, but when we begin training models at scale later, we will see the speed benefits of parallelizing our computations using CUDA. By training our models in parallel, we will be able to reduce the time this takes by a considerable amount.

Comparing PyTorch to other deep learning frameworks

PyTorch is one of the main frameworks used in deep learning today. There are other widely used frameworks available too, such as TensorFlow, Theano, and Caffe. While these are very similar in many ways, there are some key differences in how they operate. These include the following:

- How the models are computed

- The way in which the computational graphs are compiled

- The ability to create dynamic computational graphs with variable layers

- Differences in syntax

Arguably, the main difference between PyTorch and other frameworks is in the way that the models themselves are computed. PyTorch uses an automatic differentiation method called **autograd**, which allows computational graphs to be defined and executed dynamically. This is in contrast to other frameworks such as TensorFlow, which is a static framework. In these static frameworks, computational graphs must be defined and compiled before finally being executed. While using pre-compiled models may lead to efficient implementations in production, they do not offer the same level of flexibility in research and explorational projects.

Frameworks such as PyTorch do not need to pre-compile computational graphs before the model can be trained. The dynamic computational graphs used by PyTorch mean that graphs are compiled as they are executed, which allows graphs to be defined on the go. The dynamic approach to model construction is particularly useful in the field of NLP. Let's consider two sentences that we wish to perform sentiment analysis on:

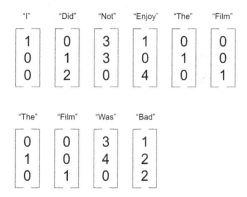

Figure 2.10 – Model construction in PyTorch

We can represent each of these sentences as a sequence of individual word vectors, which would then form our input to our neural network. However, as we can see, each of our inputs is of a different size. Within a fixed computation graph, these varying input sizes could be a problem, but for frameworks like PyTorch, models are able to adjust dynamically to account for the variation in input structure. This is one reason why PyTorch is often preferred for NLP-related deep learning.

Another major difference between PyTorch and other deep learning frameworks is syntax. PyTorch is often preferred by developers with experience in Python as it is considered to be very Pythonic in nature. PyTorch integrates well with other aspects of the Python ecosystem and it is very easy to learn if you have prior knowledge of Python. We will demonstrate PyTorch syntax now by coding up our own neural network from scratch.

Building a simple neural network in PyTorch

We will now walk through building a neural network from scratch in PyTorch. Here, we have a small `.csv` file containing several examples of images from the MNIST dataset. The MNIST dataset consists of a collection of hand-drawn digits between 0 and 9 that we want to attempt to classify. The following is an example from the MNIST dataset, consisting of a hand-drawn digit 1:

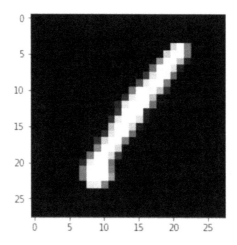

Figure 2.11 – Sample image from the MNIST dataset

These images are 28x28 in size: 784 pixels in total. Our dataset in `train.csv` consists of 1,000 of these images, with each consisting of 784 pixel values, as well as the correct classification of the digit (in this case, 1).

Loading the data

We will begin by loading the data, as follows:

1. First, we need to load our training dataset, as follows:

    ```
    train = pd.read_csv("train.csv")
    train_labels = train['label'].values
    train = train.drop("label",axis=1).values.
    reshape(len(train),1,28,28)
    ```

    Notice that we reshaped our input to (1, 1, 28, 28), which is a tensor of 1,000 images, each consisting of 28x28 pixels.

2. Next, we convert our training data and training labels into PyTorch tensors so they can be fed into the neural network:

    ```
    X = torch.Tensor(train.astype(float))
    y = torch.Tensor(train_labels).long()
    ```

Note the data types of these two tensors. A float tensor comprises 32-bit floating-point numbers, while a long tensor consists of 64-bit integers. Our X features must be floats in order for PyTorch to be able to compute gradients, while our labels must be integers within this classification model (as we're trying to predict values of 1, 2, 3, and so on), so a prediction of 1.5 wouldn't make sense.

Building the classifier

Next, we can start to construct our actual neural network classifier:

```
class MNISTClassifier(nn.Module):
    def __init__(self):
        super().__init__()
        self.fc1 = nn.Linear(784, 392)
        self.fc2 = nn.Linear(392, 196)
        self.fc3 = nn.Linear(196, 98)
        self.fc4 = nn.Linear(98, 10)
```

We build our classifier as if we were building a normal class in Python, inheriting from nn.Module in PyTorch. Within our init method, we define each of the layers of our neural network. Here, we define fully connected linear layers of varying sizes.

Our first layer takes **784** inputs as this is the size of each of our images to classify (28x28). We then see that the output of one layer must have the same value as the input of the next one, which means our first fully connected layer outputs **392** units and our second layer takes **392** units as input. This is repeated for each layer, with them having half the number of units each time until we reach our final fully connected layer, which outputs **10** units. This is the length of our classification layer.

Our network now looks something like this:

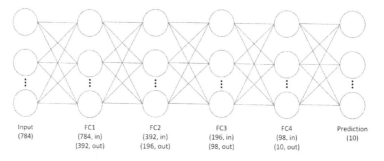

Figure 2.12 – Our neural network

Here, we can see that our final layer outputs **10** units. This is because we wish to predict whether each image is a digit between 0 and 9, which is 10 different possible classifications in total. Our output is a vector of length **10** and contains predictions for each of the 10 possible values of the image. When making a final classification, we take the digit classification that has the highest value as the model's final prediction. For example, for a given prediction, our model might predict the image is type 1 with a probability of 10%, type 2 with a probability of 10%, and type 3 with a probability of 80%. We would, therefore, take type 3 as the prediction as it was predicted with the highest probability.

Implementing dropout

Within the `init` method of our `MNISTClassifier` class, we also define a dropout method in order to help regularize the network:

```
self.dropout = nn.Dropout(p=0.2)
```

Dropout is a way of regularizing our neural networks to prevent overfitting. On each training epoch, for each node in a layer that has dropout applied, there is a probability (here, defined as $p = 20\%$) that each node within the layer will not be used in training/ backpropagation. This means that when training, our network becomes robust toward overfitting since each node will not be used in every iteration of the training process. This prevents our network from becoming too reliant on predictions from specific nodes within our network.

Defining the forward pass

Next, we define the forward pass within our classifier:

```
def forward(self, x):
    x = x.view(x.shape[0], -1)
    x = self.dropout(F.relu(self.fc1(x)))
    x = self.dropout(F.relu(self.fc2(x)))
    x = self.dropout(F.relu(self.fc3(x)))
    x = F.log_softmax(self.fc4(x), dim=1)
```

The `forward()` method within our classifier is where we apply our activation functions and define where dropout is applied within our network. Our `forward` method defines the path our input will take through the network. It first takes our input, x, and reshapes it for use within the network, transforming it into a one-dimensional vector. We then pass it through our first fully connected layer and wrap it in a `ReLU` activation function to make it non-linear. We also wrap it in our dropout, as defined in our `init` method. We repeat this process for all the other layers in the network.

For our final prediction layer, we wrap it in a log `softmax` layer. We will use this to easily calculate our loss function, as we will see next.

Setting the model parameters

Next, we define our model parameters:

```
model = MNISTClassifier()
loss_function = nn.NLLLoss()
opt = optim.Adam(model.parameters(), lr=0.001)
```

We initialize an instance of our `MNISTClassifier` class as a model. We also define our loss as a **Negative Log Likelihood Loss**:

$$Loss(y) = -log(y)$$

Let's assume our image is of a number 7. If we predict class 7 with probability 1, our loss will be -*log(1)* = *0*, but if we only predict class 7 with probability 0.7, our loss will be -*log(0.7)* = *0.3*. This means that our loss approaches infinity the further away from the correct prediction we are:

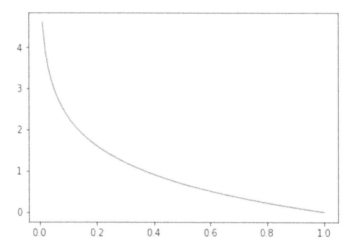

Figure 2.13 – Representation of loss for our network

This is then summed over all the correct classes in our dataset to compute the total loss. Note that we defined a log softmax when building the classifier as this already applies a softmax function (restricting the predicted output to be between 0 and 1) and takes the log. This means that *log(y)* is already calculated, so all we need to do to compute the total loss on the network is calculate the negative sum of the outputs.

We will also define our optimizer as an Adam optimizer. An optimizer controls the **learning rate** within our model. The learning rate of a model defines how big the parameter updates are during each epoch of training. The larger the size of the learning rate, the larger the size of the parameter updates during gradient descent. An optimizer dynamically controls this learning rate so that when a model is initialized, the parameter updates are large. However, as the model learns and moves closer to the point where loss is minimized, the optimizer controls the learning rate, so the parameter updates become smaller and the local minimum can be located more precisely.

Training our network

Finally, we can actually start training our network:

1. First, create a loop that runs once for each epoch of our training. Here, we will run our training loop for 50 epochs. We first take our input tensor of images and our output tensor of labels and transform them into PyTorch variables. A variable is a PyTorch object that contains a backward() method that we can use to perform backpropagation through our network:

```
for epoch in range(50):
    images = Variable(X)
    labels = Variable(y)
```

2. Next, we call zero_grad() on our optimizer to set our calculated gradients to zero. Within PyTorch, gradients are calculated cumulatively on each backpropagation. While this is useful in some models, such as when training RNNs, for our example, we wish to calculate the gradients from scratch after each epoch, so we make sure to reset the gradients to zero after each pass:

```
opt.zero_grad()
```

3. Next, we use our model's current state to make predictions on our dataset. This is effectively our forward pass as we then use these predictions to calculate our loss:

```
outputs = model(images)
```

4. Using the outputs and the true labels of our dataset, we calculate the total loss of our model using the defined loss function, which in this case is the negative log likelihood. On calculating this loss, we can then make a backward() call to backpropagate our loss through the network. We then use step() using our optimizer in order to update our model parameters accordingly:

```
loss = loss_function(outputs, labels)
loss.backward()
opt.step()
```

5. Finally, after each epoch is complete, we print the total loss. We can observe this to make sure our model is learning:

```
print ('Epoch [%d/%d] Loss: %.4f' %(epoch+1, 50,
        loss.data.item()))
```

In general, we would expect the loss to decrease after every epoch. Our output will look something like this:

```
Epoch [1/50] Loss: 9.7278
Epoch [2/50] Loss: 6.6400
Epoch [3/50] Loss: 4.6651
Epoch [4/50] Loss: 3.4805
Epoch [5/50] Loss: 2.7262
```

Figure 2.14 – Training epochs

Making predictions

Now that our model has been trained, we can use this to make predictions on unseen data. We begin by reading in our test set of data (which was not used to train our model):

```
test = pd.read_csv("test.csv")
test_labels = test['label'].values
test = test.drop("label",axis=1).values.reshape(len(test),
          1,28,28)


X_test = torch.Tensor(test.astype(float))
y_test = torch.Tensor(test_labels).long()
```

Here, we perform the same steps we performed when we loaded our training set of data: we reshape our test data and transform it into PyTorch tensors. Next, to predict using our trained model, we simply run the following command:

```
preds = model(X_test)
```

In the same way that we calculated our outputs on the forward pass of our training data in our model, we now pass our test data through the model and obtain predictions. We can view the predictions for one of the images like so:

```
print(preds[0])
```

This results in the following output:

```
tensor([-5.8550, -6.8315, -7.0251, -7.9013, -4.3035, -6.1763, -6.9913, -6.0805,
        -5.7120, -0.0277], grad_fn=<SelectBackward>)
```

Figure 2.15 – Prediction outputs

Here, we can see that our prediction is a vector of length 10, with a prediction for each of the possible classes (digits between 0 and 9). The one with the highest predicted value is the one our model chooses as its prediction. In this case, it is the 10th unit of our vector, which equates to the digit 9. Note that since we used log softmax earlier, our predictions are logs and not raw probabilities. To convert these back into probabilities, we can just transform them using *x*.

We can now construct a summary DataFrame containing our true test data labels, as well as the labels our model predicted:

```
_, predictionlabel = torch.max(preds.data, 1)
predictionlabel = predictionlabel.tolist()

predictionlabel = pd.Series(predictionlabel)
test_labels = pd.Series(test_labels)

pred_table = pd.concat([predictionlabel, test_labels], axis=1)
pred_table.columns = ['Predicted Value', 'True Value']

display(pred_table.head())
```

This results in the following output:

	Predicted Value	True Value
0	9	9
1	8	5
2	2	2
3	4	4
4	1	1

Figure 2.16 – Prediction table

Note how the `torch.max()` function automatically selects the prediction with the highest value. We can see here that, based on a small selection of our data, our model appears to be making some good predictions!

Evaluating our model

Now that we have some predictions from our model, we can use these predictions to evaluate how good our model is. One rudimentary way of evaluating model performance is **accuracy**, as discussed in the previous chapter. Here, we simply calculate our correct predictions (where the predicted image label is equal to the actual image label) as a percentage of the total number of predictions our model made:

```
preds = len(predictionlabel)
correct = len([1 for x,y in zip(predictionlabel, test_labels)
               if x==y])
print((correct/preds)*100)
```

This results in the following output:

89.5

Figure 2.17 – Accuracy score

Congratulations! Your first neural network was able to correctly identify almost 90% of unseen digit images. As we progress, we will see that there are more sophisticated models that may lead to improved performance. However, for now, we have demonstrated that creating a simple deep neural network is very simple using PyTorch. This can be coded up in just a few lines and leads to performance above and beyond what is possible with basic machine learning models such as regression.

NLP for PyTorch

Now that we have learned how to build neural networks, we will see how it is possible to build models for NLP using PyTorch. In this example, we will create a basic bag-of-words classifier in order to classify the language of a given sentence.

Setting up the classifier

For this example, we'll take a selection of sentences in Spanish and English:

1. First, we split each sentence into a list of words and take the language of each sentence as a label. We take a section of sentences to train our model on and keep a small section to one side as our test set. We do this so that we can evaluate the performance of our model after it has been trained:

```
("This is my favourite chapter".lower().split(),\
 "English"),
("Estoy en la biblioteca".lower().split(), "Spanish")
```

Note that we also transform each word into lowercase, which stops words being double counted in our bag-of-words. If we have the word book and the word Book, we want these to be counted as the same word, so we transform these into lowercase.

2. Next, we build our word index, which is simply a dictionary of all the words in our corpus, and then create a unique index value for each word. This can be easily done with a short for loop:

```
word_dict = {}
i = 0
for words, language in training_data + test_data:
    for word in words:
        if word not in word_dict:
            word_dict[word] = i
            i += 1
print(word_dict)
```

This results in the following output:

```
{'the': 5, 'libro': 22, 'a': 9, 'estoy': 16, 'am': 7, 'paginas': 1, 'chap
ter': 15, 'library': 6, 'la': 18, 'un': 21, 'favourite': 14, 'reading':
8, 'will': 3, 'i': 2, 'leyendo': 23, 'en': 17, 'not': 24, 'visit': 4, 'bo
ok': 10, 'is': 12, 'biblioteca': 19, 'this': 11, 'my': 13, 'tengo': 20,
'veinte': 0}
```

Figure 2.18 – Setting up the classifier

Note that here, we looped through all our training data and test data. If we just created our word index on training data, when it came to evaluating our test set, we would have new words that were not seen in the original training, so we wouldn't be able to create a true bag-of-words representation for these words.

3. Now, we build our classifier in a similar fashion to how we built our neural network in the previous section; that is, by building a new class that inherits from `nn.Module`.

   Here, we define our classifier so that it consists of a single linear layer with log softmax activation functions approximating a logistic regression. We could easily extend this to operate as a neural network by adding extra linear layers here, but a single layer of parameters will serve our purpose. Pay close attention to the input and output sizes of our linear layer:

```
corpus_size = len(word_dict)
languages = 2
label_index = {"Spanish": 0, "English": 1}


class BagofWordsClassifier(nn.Module):

    def __init__(self, languages, corpus_size):
        super(BagofWordsClassifier, self).__init__()
        self.linear = nn.Linear(corpus_size, languages)

    def forward(self, bow_vec):
        return F.log_softmax(self.linear(bow_vec), dim=1)
```

The input is of length `corpus_size`, which is just the total count of unique words in our corpus. This is because each input to our model will be a bag-of-words representation, consisting of the counts of words in each sentence, with a count of 0 if a given word does not appear in our sentence. Our output is of size 2, which is our number of languages to predict. Our final predictions will consist of a probability that our sentence is English versus the probability that our sentence is Spanish, with our final prediction being the one with the highest probability.

4. Next, we define some utility functions. We first define make_bow_vector, which takes the sentence and transforms it into a bag-of-words representation. We first create a vector consisting of all zeros. We then loop through them and for each word in the sentence, we increment the count of that index within the bag-of-words vector by one. We finally reshape this vector using with .view() for entry into our classifier:

```
def make_bow_vector(sentence, word_index):
    word_vec = torch.zeros(corpus_size)
    for word in sentence:
        word_vec[word_dict[word]] += 1
    return word_vec.view(1, -1)
```

5. Similarly, we define make_target, which simply takes the label of the sentence (Spanish or English) and returns its relevant index (0 or 1):

```
def make_target(label, label_index):
    return torch.LongTensor([label_index[label]])
```

6. We can now create an instance of our model, ready for training. We also define our loss function as Negative Log Likelihood as we are using a log softmax function, and then define our optimizer in order to use standard **stochastic gradient descent (SGD)**:

```
model = BagofWordsClassifier(languages, corpus_size)
loss_function = nn.NLLLoss()
optimizer = optim.SGD(model.parameters(), lr=0.1)
```

Now, we are ready to train our model.

Training the classifier

First, we set up a loop consisting of the number of epochs we wish our model to run for. In this instance, we will select 100 epochs.

Within this loop, we first zero our gradients (as otherwise, PyTorch calculates gradients cumulatively) and then for each sentence/label pair, we transform each into a bag-of-words vector and target, respectively. We then calculate the predicted output of this particular sentence pair by making a forward pass of our data through the current state of our model.

Using this prediction, we then take our predicted and actual labels and call our defined `loss_function` on the two to obtain a measure of loss for this sentence. By calling `backward()`, we then backpropagate this loss through our model and by calling `step()` on our optimizer, we update our model parameters. Finally, we print our loss after every 10 training steps:

```
for epoch in range(100):
    for sentence, label in training_data:

        model.zero_grad()

        bow_vec = make_bow_vector(sentence, word_dict)
        target = make_target(label, label_index)

        log_probs = model(bow_vec)

        loss = loss_function(log_probs, target)
        loss.backward()
        optimizer.step()

    if epoch % 10 == 0:
        print('Epoch: ',str(epoch+1),', Loss: ' +
                      str(loss.item()))
```

This results in the following output:

```
Epoch:  1 , Loss: 0.013411426916718483
Epoch:  11 , Loss: 0.012176347896456718
Epoch:  21 , Loss: 0.011149131692945957
Epoch:  31 , Loss: 0.0102814557030797
Epoch:  41 , Loss: 0.009538905695080757
Epoch:  51 , Loss: 0.00889623910188675
Epoch:  61 , Loss: 0.00833461619913578
Epoch:  71 , Loss: 0.007839507423341274
Epoch:  81 , Loss: 0.007399887777864933
Epoch:  91 , Loss: 0.007006953936070204
```

Figure 2.19 – Training loss

Here, we can see that our loss is decreasing over time as our model learns. Although our training set in this example is very small, we can still demonstrate that our model has learned something useful, as follows:

1. We evaluate our model on a couple of sentences from our test data that our model was not trained on. Here, we first set `torch.no_grad()`, which deactivates the `autograd` engine as there is no longer any need to calculate gradients as we are no longer training our model. Next, we take our test sentence and transform it into a bag-of-words vector and feed it into our model to obtain predictions.

2. We then simply print the sentence, the true label of the sentence, and then the predicted probabilities. Note that we transform the predicted values from log probabilities back into probabilities. We obtain two probabilities for each prediction, but if we refer back to the label index, we can see that the first probability (index 0) corresponds to Spanish, whereas the other one corresponds to English:

```python
def make_predictions(data):

    with torch.no_grad():
        sentence = data[0]
        label = data[1]
        bow_vec = make_bow_vector(sentence, word_dict)
        log_probs = model(bow_vec)
        print(sentence)
        print(label + ':')
        print(np.exp(log_probs))


make_predictions(test_data[0])
make_predictions(test_data[1])
```

This results in the following output:

```
['estoy', 'leyendo']
Spanish:
tensor([[0.8606, 0.1394]])
['this', 'is', 'not', 'my', 'favourite', 'book']
English:
tensor([[0.0074, 0.9926]])
```

Figure 2.20 – Predicted output

Here, we can see that for both our predictions, our model predicts the correct answer, but why is this? What exactly has our model learned? We can see that our first test sentence contains the word estoy, which was previously seen in a Spanish sentence within our training set. Similarly, we can see that the word book was seen within our training set in an English sentence. Since our model consists of a single layer, the parameters on each of our nodes are easy to interpret.

3.   Here, we define a function that takes a word as input and returns the weights on each of the parameters within the layer. For a given word, we get the index of this word from our dictionary and then select these parameters from the same index within the model. Note that our model returns two parameters as we are making two predictions; that is, the model's contribution to the Spanish prediction and the model's contribution to the English prediction:

```
def return_params(word):
    index = word_dict[word]
    for p in model.parameters():
        dims = len(p.size())
        if dims == 2:
            print(word + ':')
            print('Spanish Parameter = ' +
                    str(p[0][index].item()))
            print('English Parameter = ' +
                    str(p[1][index].item()))
            print('\n')


return_params('estoy')
return_params('book')
```

This results in the following output:

```
book:
Spanish Parameter = -0.49991941452026367
English Parameter = 0.4414403736591339


estoy:
Spanish Parameter = 0.43009454011917114
English Parameter = -0.5826961398124695
```

Figure 2.21 – Predicted output for the updated function

Here, we can see that for the word `estoy`, this parameter is positive for the Spanish prediction and negative for the English one. This means that for each count of the word "estoy" in our sentence, the sentence becomes more likely to be a Spanish sentence. Similarly, for the word `book`, we can see that this contributes positively to the prediction that the sentence is English.

We can show that our model has only learned based on what it has been trained on. If we try to predict a word the model hasn't been trained on, we can see it is unable to make an accurate decision. In this case, our model thinks that the English word "`not`" is Spanish:

```
new_sentence = (["not"],"English")
make_predictions(new_sentence)
```

This results in the following output:

```
['not']
English:
tensor([[0.7152, 0.2848]])
```

Figure 2.22 – Final output

Summary

In this chapter, we introduced PyTorch and some of its key features. Hopefully, you now have a better understanding of how PyTorch differs from other deep learning frameworks and how it can be used to build basic neural networks. While these simple examples are just the tip of the iceberg, we have illustrated that PyTorch is an immensely powerful tool for NLP analysis and learning.

In future chapters, we will demonstrate how the unique properties of PyTorch can be utilized to build highly sophisticated models for solving very complex machine learning tasks.

Section 2: Fundamentals of Natural Language Processing

In this section, you will learn about the fundamentals of building a **Natural Language Processing (NLP)** application. You will also learn how to use various NLP techniques, such as word embeddings, CBOW, and tokenization in PyTorch in this section.

This section contains the following chapters:

- *Chapter 3, NLP and Text Embeddings*
- *Chapter 4, Stemming and Lemmatization*

3
NLP and Text Embeddings

There are many different ways of representing text in deep learning. While we have covered basic **bag-of-words** (**BoW**) representations, unsurprisingly, there is a far more sophisticated way of representing text data known as embeddings. While a BoW vector acts only as a count of words within a sentence, embeddings help to numerically define the actual meaning of certain words.

In this chapter, we will explore text embeddings and learn how to create embeddings using a continuous BoW model. We will then move on to discuss n-grams and how they can be used within models. We will also cover various ways in which tagging, chunking, and tokenization can be used to split up NLP into its various constituent parts. Finally, we will look at TF-IDF language models and how they can be useful in weighting our models toward infrequently occurring words.

The following topics will be covered in the chapter:

- Word embeddings
- Exploring CBOW
- Exploring n-grams
- Tokenization

- Tagging and chunking for parts of speech
- TF-IDF

Technical requirements

GLoVe vectors can be downloaded from `https://nlp.stanford.edu/projects/glove/`. It is recommended to use the `glove.6B.50d.txt` file as it is much smaller than the other files and will be much faster to process. NLTK will be required for later parts of this chapter. All the code for this chapter can be found at `https://github.com/PacktPublishing/Hands-On-Natural-Language-Processing-with-PyTorch-1.x`.

Embeddings for NLP

Words do not have a natural way of representing their meaning. In images, we already have representations in rich vectors (containing the values of each pixel within the image), so it would clearly be beneficial to have a similarly rich vector representation of words. When parts of language are represented in a high-dimensional vector format, they are known as **embeddings**. Through analysis of a corpus of words, and by determining which words appear frequently together, we can obtain an n-length vector for each word, which better represents the semantic relationship of each word to all other words. We saw previously that we can easily represent words as one-hot encoded vectors:

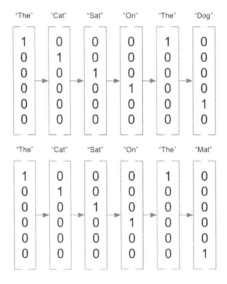

Figure 3.1 – One-hot encoded vectors

On the other hand, embeddings are vectors of length n (in the following example, $n = 3$) that can take any value:

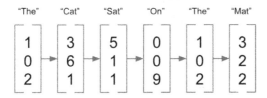

Figure 3.2 – Vectors with n=3

These embeddings represent the word's vector in n-dimensional space (where n is the length of the embedding vectors), and words with similar vectors within this space are considered to be more similar in meaning. While these embeddings can be of any size, they are generally of much lower dimensionality than the BoW representation. The BOW representation requires vectors that are of the length of the entire corpus, which, when looking at a whole language, could become very large very quickly. Although embeddings are of a high enough dimensionality to represent the individual words, they are generally not much larger than a few hundred dimensions. Furthermore, the BOW vectors are generally very sparse, consisting mostly of zeros, whereas embeddings are rich in data and every dimension contributes to the overall representation of the word. The lower dimensionality and the fact that they are not sparse makes performing deep learning on embeddings much more efficient than performing it on BOW representations.

GLoVe

We can download a set of pre-calculated word embeddings to demonstrate how they work. For this, we will use **Global Vectors for Word Representation (GLoVe)** embeddings, which can be downloaded from here: `https://nlp.stanford.edu/projects/glove/`. These embeddings are calculated on a very large corpus of NLP data and are trained on a word co-occurrence matrix. This is based on the notion that words that appear together are more likely to have similar meaning. For instance, the word *sun* is likely to appear more frequently with the word *hot* as opposed to the word *cold*, so it is more likely that *sun* and *hot* are considered more similar.

We can validate that this is true by examining the individual GLoVe vectors:

1. We first create a simple function to load our GLoVe vectors from a text file. This just builds a dictionary where the index is each word in the corpus and the values are the embedding vector:

```python
def loadGlove(path):
    file = open(path,'r')
    model = {}
    for l in file:
        line = l.split()
        word = line[0]
        value = np.array([float(val) for val in
                          line[1:]])
        model[word] = value
    return model


glove = loadGlove('glove.6B.50d.txt')
```

2. This means we can access a single vector by just calling it from the dictionary:

```python
glove['python']
```

This results in the following output:

```
array([ 0.5897  , -0.55043 , -1.0106  ,  0.41226 ,  0.57348 ,  0.23464 ,
       -0.35773 , -1.78    ,  0.10745 ,  0.74913 ,  0.45013 ,  1.0351  ,
        0.48348 ,  0.47954 ,  0.51908 , -0.15053 ,  0.32474 ,  1.0789  ,
       -0.90894 ,  0.42943 , -0.56388 ,  0.69961 ,  0.13501 ,  0.16557 ,
       -0.063592,  0.35435 ,  0.42819 ,  0.1536  , -0.47018 , -1.0935  ,
        1.361   , -0.80821 , -0.674   ,  1.2606  ,  0.29554 ,  1.0835  ,
        0.2444  , -1.1877  , -0.60203 , -0.068315,  0.66256 ,  0.45336 ,
       -1.0178  ,  0.68267 , -0.20788 , -0.73393 ,  1.2597  ,  0.15425 ,
       -0.93256 , -0.15025 ])
```

Figure 3.3 – Vector output

We can see that this returns a 50-dimensional vector embedding for the word Python. We will now introduce the concept of **cosine similarity** to compare how similar two vectors are. Vectors will have a similarity of 1 if the angle in the *n*-dimensional space between them is 0 degrees. Values with high cosine similarity can be considered similar, even if they are not equal. This can be calculated using the following formula, where A and B are the two embedding vectors being compared:

$$\frac{\sum A \cdot B}{\sqrt{\sum A^2} \times \sqrt{\sum B^2}}$$

3. We can calculate this easily in Python using the cosine_similarity() function from Sklearn. We can see that cat and dog have similar vectors as they are both animals:

```
cosine_similarity(glove['cat'].reshape(1, -1),
glove['dog'].reshape(1, -1))
```

This results in the following output:

array([[0.92180053]])

Figure 3.4 – Cosine similarity output for cat and dog

4. However, cat and piano are quite dissimilar as they are two seemingly unrelated items:

```
cosine_similarity(glove['cat'].reshape(1, -1),
glove['piano'].reshape(1, -1))
```

This results in the following output:

array([[0.19825255]])

Figure 3.5 – Cosine similarity output for cat and piano

Embedding operations

Since embeddings are vectors, we can perform operations on them. For example, let's say we take the embeddings for the following sorts and we calculate the following:

Queen-Woman+Man

With this, we can approximate the embedding for *king*. This essentially replaces the *Woman* vector component from *Queen* with the *Man* vector to arrive at this approximation. We can graphically illustrate this as follows:

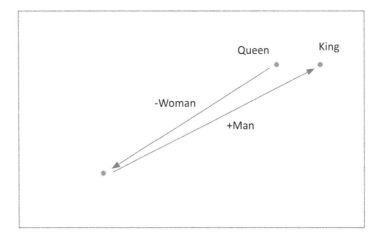

Figure 3.6 – Graphical representation of the example

Note that in this example, we illustrate this graphically in two dimensions. In the case of our embeddings, this is happening in a 50-dimensional space. While this is not exact, we can verify that our calculated vector is indeed similar to the GLoVe vector for **King**:

```
predicted_king_embedding = glove['queen'] - glove['woman'] +
glove['man']
cosine_similarity(predicted_king_embedding.reshape(1, -1),
glove['king'].reshape(1, -1))
```

This results in the following output:

```
array([[0.85888392]])
```

Figure 3.7 – Output for the GLoVe vector

While GLoVe embeddings are very useful pre-calculated embeddings, it is actually possible for us to calculate our own embeddings. This may be useful when we are analyzing a particularly unique corpus. For example, the language used on Twitter may differ from the language used on Wikipedia, so embeddings trained on one may not be useful for the other. We will now demonstrate how we can calculate our own embeddings using a continuous bag-of-words.

Exploring CBOW

The **continuous bag-of-words (CBOW)** model forms part of Word2Vec – a model created by Google in order to obtain vector representations of words. By running these models over a very large corpus, we are able to obtain detailed representations of words that represent their semantic and contextual similarity to one another. The Word2Vec model consists of two main components:

- **CBOW**: This model attempts to predict the target word in a document, given the surrounding words.

- **Skip-gram**: This is the opposite of CBOW; this model attempts to predict the surrounding words, given the target word.

Since these models perform similar tasks, we will focus on just one for now, specifically CBOW. This model aims to predict a word (the **target word**), given the other words around it (known as the **context** words). One way of accounting for context words could be as simple as using the word directly before the target word in the sentence to predict the target word, whereas more complex models could use several words before and after the target word. Consider the following sentence:

PyTorch is a deep learning framework

Let's say we want to predict the word *deep*, given the context words:

PyTorch is a {target_word} learning framework

We could look at this in a number of ways:

	Context Words	Representation
Previous Word	"a"	$n - 1$
Next Word	"learning"	$n + 1$
Previous and Next (Window length = 1)	"a", "learning"	$n - 1, n + 1$
Previous and Next (Window length = 2)	"is","a","learning","framework"	$n - 2, n - 1, n + 1, n + 2$

Figure 3.8 – Table of context and representations

For our CBOW model, we will use a window of length 2, which means for our model's (*X, y*) input/output pairs, we use *([n-2, n-1, n+1, n+2, n])*, where *n* is our target word being predicted.

Using these as our model inputs, we will train a model that includes an embedding layer. This embedding layer automatically forms an *n*-dimensional representation of the words in our corpus. However, to begin with, this layer is initialized with random weights. These parameters are what will be learned using our model so that after our model has finished training, this embedding layer can be used can be used to encode our corpus in an embedded vector representation.

CBOW architecture

We will now design the architecture of our model in order to learn our embeddings. Here, our model takes an input of four words (two before our target word and two after) and trains it against an output (our target word). The following representation is an illustration of how this might look:

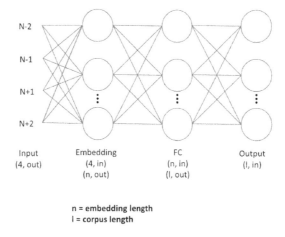

Figure 3.9 – CBOW architecture

Our input words are first fed through an embedding layer, represented as a tensor of size (n,l), where n is the specified length of our embeddings and l is the number of words in our corpus. This is because every word within the corpus has its own unique tensor representation.

Using our combined (summed) embeddings from our four context words, this is then fed into a fully connected layer in order to learn the final classification of our target word against our embedded representation of our context words. Note that our predicted/target word is encoded as a vector that's the length of our corpus. This is because our model effectively predicts the probability of each word in the corpus to be the target word, and the final classification is the one with the highest probability. We then obtain a loss, backpropagate this through our network, and update the parameters on the fully connected layer, as well as the embeddings themselves.

The reason this methodology works is because our learned embeddings represent semantic similarity. Let's say we train our model on the following:

X = ["is", "a", "learning", "framework"]; y = "deep"

What our model is essentially learning is that the combined embedding representation of our target words is semantically similar to our target word. If we repeat this over a large enough corpus of words, we will find that our word embeddings begin to resemble our previously seen GLoVe embeddings, where semantically similar words appear to one another within the embedding space.

Building CBOW

We will now run through building a CBOW model from scratch, thereby demonstrating how our embedding vectors can be learned:

1. We first define some text and perform some basic text cleaning, removing basic punctuation and converting it all into lowercase:

    ```
    text = text.replace(',','').replace('.','').lower().
                             split()
    ```

2. We start by defining our corpus and its length:

    ```
    corpus = set(text)
    corpus_length = len(corpus)
    ```

3. Note that we use a set instead of a list as we are only concerned with the unique words within our text. We then build our corpus index and our inverse corpus index. Our corpus index will allow us to obtain the index of a word given the word itself, which will be useful when encoding our words for entry into our network. Our inverse corpus index allows us to obtain a word, given the index value, which will be used to convert our predictions back into words:

    ```
    word_dict = {}
    inverse_word_dict = {}

    for i, word in enumerate(corpus):
        word_dict[word] = i
        inverse_word_dict[i] = word
    ```

4. Next, we encode our data. We loop through our corpus and for each target word, we capture the context words (the two words before and the two words after). We append this with the target word itself to our dataset. Note how we begin this process from the third word in our corpus (index = 2) and stop it two steps before the end of the corpus. This is because the two words at the beginning won't have two words before them and, similarly, the two words at the end won't have two words after them:

```
data = []

for i in range(2, len(text) - 2):
    sentence = [text[i-2], text[i-1],
                text[i+1], text[i+2]]
    target = text[i]
    data.append((sentence, target))

print(data[3])
```

This results in the following output:

```
(['haunted', 'my', 'i', 'need'], 'dreams')
```

Figure 3.10 – Encoding the data

5. We then define the length of our embeddings. While this can technically be any number you wish, there are some tradeoffs to consider. While higher-dimensional embeddings can lead to a more detailed representation of the words, the feature space also becomes sparser, which means high-dimensional embeddings are only appropriate for large corpuses. Furthermore, larger embeddings mean more parameters to learn, so increasing the embedding size can increase training time significantly. We are only training on a very small dataset, so we have opted to use embeddings of size 20:

```
embedding_length = 20
```

Next, we define our CBOW model in PyTorch. We define our embeddings layer so that it takes a vector of corpus length in and outputs a single embedding. We define our linear layer as a fully connected layer that takes an embedding in and outputs a vector of 64. We define our final layer as a classification layer that is the same length as our text corpus.

6. We define our forward pass by obtaining and summing the embeddings for all
    input context words. This then passes through the fully connected layer with ReLU
    activation functions and finally into the classification layer, which predicts which
    word in the corpus corresponds to the summed embeddings of the context words
    the most:

```python
class CBOW(torch.nn.Module):

    def __init__(self, corpus_length, embedding_dim):
        super(CBOW, self).__init__()

        self.embeddings = nn.Embedding(corpus_length,
                                       embedding_dim)

        self.linear1 = nn.Linear(embedding_dim, 64)
        self.linear2 = nn.Linear(64, corpus_length)

        self.activation_function1 = nn.ReLU()
        self.activation_function2 = nn.LogSoftmax
                                        (dim = -1)

    def forward(self, inputs):
        embeds = sum(self.embeddings(inputs)).view(1,-1)
        out = self.linear1(embeds)
        out = self.activation_function1(out)
        out = self.linear2(out)
        out = self.activation_function2(out)
        return out
```

7. We can also define a get_word_embedding() function, which will allow
    us to extract embeddings for a given word after our model has been trained:

```python
def get_word_emdedding(self, word):
    word = torch.LongTensor([word_dict[word]])
    return self.embeddings(word).view(1,-1)
```

8.   Now, we are ready to train our model. We first create an instance of our model and define the loss function and optimizer:

```
model = CBOW(corpus_length, embedding_length)

loss_function = nn.NLLLoss()
optimizer = torch.optim.SGD(model.parameters(), lr=0.01)
```

9.   We then create a helper function that takes our input context words, gets the word indexes for each of these, and transforms them into a tensor of length 4, which forms the input to our neural network:

```
def make_sentence_vector(sentence, word_dict):
    idxs = [word_dict[w] for w in sentence]
    return torch.tensor(idxs, dtype=torch.long)

print(make_sentence_vector(['stormy','nights','when','
the'], word_dict))
```

This results in the following output:

```
tensor([ 9, 48, 43, 70])
```

Figure 3.11 – Tensor value

10. Now, we train our network. We loop through 100 epochs and for each pass, we loop through all our context words, that is, target word pairs. For each of these pairs, we load the context sentence using make_sentence_vector() and use our current model state to obtain predictions. We evaluate these predictions against our actual target in order to obtain our loss. We backpropagate to calculate the gradients and step through our optimizer to update the weights. Finally, we sum all our losses for the epoch and print this out. Here, we can see that our loss is decreasing, showing that our model is learning:

```
for epoch in range(100):
    epoch_loss = 0
    for sentence, target in data:
        model.zero_grad()
        sentence_vector = make_sentence_vector
                            (sentence, word_dict)
        log_probs = model(sentence_vector)
        loss = loss_function(log_probs, torch.tensor(
```

```
        [word_dict[target]], dtype=torch.long))
        loss.backward()
        optimizer.step()
        epoch_loss += loss.data
    print('Epoch: '+str(epoch)+', Loss: ' + str(epoch_
loss.item()))
```

This results in the following output:

```
Epoch: 0, Loss: 540.2003173828125
Epoch: 1, Loss: 479.9447937011719
Epoch: 2, Loss: 435.550537109375
Epoch: 3, Loss: 396.0237731933594
Epoch: 4, Loss: 357.2877502441406
Epoch: 5, Loss: 318.30047607421875
```

Figure 3.12 – Training our network

Now that our model has been trained, we can make predictions. We define a couple of functions to allow us to do so. get_predicted_result() returns the predicted word from the array of predictions, while our predict_sentence() function makes a prediction based on the context words.

11. We split our sentences into individual words and transform them into an input vector. We then create our prediction array by feeding this into our model and get our final predicted word by using the get_predicted_result() function. We also print the two words before and after the predicted target word for context. We can run a couple of predictions to validate our model is working correctly:

```
def get_predicted_result(input, inverse_word_dict):
    index = np.argmax(input)
    return inverse_word_dict[index]


def predict_sentence(sentence):
    sentence_split = sentence.replace('.','').lower().
                            split()
    sentence_vector = make_sentence_vector(sentence_
                        split, word_dict)
    prediction_array = model(sentence_vector).data.
                            numpy()
    print('Preceding Words: {}\n'.format(sentence_
            split[:2]))
```

```
        print('Predicted Word: {}\n'.format(get_predicted_
            result(prediction_array[0], inverse_
            word_dict)))
        print('Following Words: {}\n'.format(sentence_
            split[2:]))


predict_sentence('to see leap and')
```

This results in the following output:

```
Preceding Words: ['to', 'see']

Predicted Word: him

Following Words: ['leap', 'and']
```

Figure 3.13 – Predicted values

12. Now that we have a trained model, we are able to use the `get_word_embedding()` function in order to return the 20 dimensions word embedding for any word in our corpus. If we needed our embeddings for another NLP task, we could actually extract the weights from the whole embedding layer and use this in our new model:

```
print(model.get_word_emdedding('leap'))
```

This results in the following output:

```
tensor([[ 8.6251e-01, -3.6810e-01, -1.1397e+00, -4.2561e-02,  3.1203e-01,
         -6.3440e-01,  1.0768e+00,  2.7745e-01, -5.6835e-01, -2.0671e+00,
          2.3117e-01, -5.5059e-02,  1.8441e+00, -1.7148e-01,  8.4483e-01,
         -6.5940e-02,  1.2200e+00,  1.0388e+00, -3.9075e-04,  1.3893e+00]],
       grad_fn=<ViewBackward>)
```

Figure 3.14 – Tensor value after editing the model

Here, we have demonstrated how to train a CBOW model for creating word embeddings. In reality, to create reliable embeddings for a corpus, we would require a very large dataset to be able to truly capture the semantic relationship between all the words. Because of this, it may be preferable to use pre-trained embeddings such as GLoVe, which have been trained on a very large corpus of data, for your models, but there may be cases where it would be preferable to train a brand new set of embeddings from scratch; for example, when analyzing a corpus of data that doesn't resemble normal NLP (for example, Twitter data where users may speak in short abbreviations and not use full sentences).

Exploring n-grams

In our CBOW model, we successfully showed that the meaning of the words is related to the context of the words around it. It is not only our context words that influence the meaning of words in a sentence, but the order of those words as well. Consider the following sentences:

The cat sat on the dog

The dog sat on the cat

If you were to transform these two sentences into a bag-of-words representation, we would see that they are identical. However, by reading the sentences, we know they have completely different meanings (in fact, they are the complete opposite!). This clearly demonstrates that the meaning of a sentence is not just the words it contains, but the order in which they occur. One simple way of attempting to capture the order of words within a sentence is by using n-grams.

If we perform a count on our sentences, but instead of counting individual words, we now count the distinct two-word pairings that occur within the sentences, this is known as using **bi-grams**:

	the_cat	the_dog	cat_sat	dog_sat	sat_on	on_the
Sentence 1	1	1	1	0	1	1
Sentence 2	1	1	0	1	1	1

Figure 3.15 – Tabular representation of bi-grams

We can represent this as follows:

The cat sat on the dog -> [1,1,1,0,1,1]

The dog sat on the cat -> [1,1,0,1,1,1]

These pairs of words attempt to capture the order the words appear in within a sentence, not just their frequency. Our first sentence contains the bi-gram *cat sat*, while the other one contains *dog sat*. These bigrams clearly help add much more context to our sentence than just using raw word counts.

We are not limited to pairs of words. We can also look at distinct word triplets, known as **trigrams**, or indeed any distinct number of words. We can use n-grams as inputs into our deep learning models instead of just a singular word, but when using n-gram models, it is worth noting that your feature space can become very large very quickly and may make machine learning very slow. If a dictionary contains all the words in the English language, a dictionary containing all distinct pairs of words would be several orders of magnitude larger!

N-gram language modeling

One thing that n-grams help us do is understand how natural language is formed. If we think of a language as being represented by parts of smaller word pairs (bigrams) instead of single words, we can begin to model language as a probabilistic model where the probability that a word appears in a sentence depends on the words that appeared before it.

In a **unigram** model, we assume that all the words have a finite probability of appearing based on the distribution of the words in a corpus or document. Let's take a document consisting of one sentence:

My name is my name

Based on this sentence, we can generate a distribution of words whereby each word has a given probability of occurring based on its frequency within the document:

Word	P(W)
My	0.4
Name	0.4
Is	0.2

Figure 3.16 – Tabular representation of a unigram

We could then draw words randomly from this distribution in order to generate new sentences:

Name is Name my my

But as we can see, this sentence doesn't make any sense, illustrating the problems of using a unigram model. Because the probability of each word occurring is independent of all the other words in the sentence, there is no consideration given to the order or context of the words appearing. This is where n-gram models are useful.

We will now consider using a **bigram** language model. This calculation takes the probability of a word occurring, given the word that appears before it:

$$p(W_n|W_{n-1}) = \frac{p(Wn-1, W_n)}{p(Wn-1)}$$

This means that the probability of a word occurring, given the previous word, is the probability of the word n-gram occurring divided by the probability of the previous word occurring. Let's say we are trying to predict the next word in the following sentence:

My favourite language is ___

Along with this, we're given the following n-gram and word probabilities:

Word	P(W)
Python	0.05
English	0.05
Is	0.1
(Is, Python)	0.02
(Is, English)	0.01

Figure 3.17 – Tabular representation of the probabilities

With this, we could calculate the probability of Python occurring, given the probability of the previous word *is* occurring is only 20%, whereas the probability of *English* occurring is only 10%. We could expand this model further to use a trigram or any n-gram representation of words as we deem appropriate. We have demonstrated that n-gram language modeling can be used to introduce further information about word's relationships to one another into our models, rather than naively assuming that words are independently distributed.

Tokenization

Next, we will learn about tokenization for NLP, a way of pre-processing text for entry into our models. Tokenization splits our sentences up into smaller parts. This could involve splitting a sentence up into its individual words or splitting a whole document up into individual sentences. This is an essential pre-processing step for NLP that can be done fairly simply in Python:

1. We first take a basic sentence and split this up into individual words using the **word tokenizer** in NLTK:

```
text = 'This is a single sentence.'
tokens = word_tokenize(text)
print(tokens)
```

This results in the following output:

```
['This', 'is', 'a', 'single', 'sentence', '.']
```

Figure 3.18 – Splitting the sentence

2. Note how a period (.) is considered a token as it is a part of natural language. Depending on what we want to do with the text, we may wish to keep or dispose of the punctuation:

```
no_punctuation = [word.lower() for word in tokens if
word.isalpha()]
print(no_punctuation)
```

This results in the following output:

```
['this', 'is', 'a', 'single', 'sentence']
```

Figure 3.19 – Removing the punctuation

3. We can also tokenize documents into individual sentences using the **sentence tokenizer**:

```
text = "This is the first sentence. This is the second
sentence. A document contains many sentences."
print(sent_tokenize(text))
```

This results in the following output:

```
['This is the first sentence.',
 'This is the second sentence.',
 'A document contains many sentences.']
```

Figure 3.20 – Splitting multiple sentences into single sentences

4. Alternatively, we can combine the two to split into individual sentences of words:

```
print([word_tokenize(sentence) for sentence in sent_
tokenize(text)])
```

This results in the following output:

```
[['This', 'is', 'the', 'first', 'sentence', '.'], ['This',
'is', 'the', 'second', 'sentence', '.'], ['A', 'document',
'contains', 'many', 'sentences', '.']]
```

Figure 3.21 – Splitting multiple sentences into words

5. One other optional step in the process of tokenization, which is the removal of stopwords. Stopwords are very common words that do not contribute to the overall meaning of a sentence. These include words such as *a*, I, and or. We can print a complete list from NLTK using the following code:

```
stop_words = stopwords.words('english')
print(stop_words[:20])
```

This results in the following output:

```
['i', 'me', 'my', 'myself', 'we', 'our', 'ours', 'ourselve
s', 'you', "you're", "you've", "you'll", "you'd", 'your',
'yours', 'yourself', 'yourselves', 'he', 'him', 'his']
```

Figure 3.22 – Displaying stopwords

6. We can easily remove these stopwords from our words using basic list comprehension:

```
text = 'This is a single sentence.'
tokens = [token for token in word_tokenize(text) if token
not in stop_words]
print(tokens)
```

This results in the following output:

```
['This', 'single', 'sentence', '.']
```

Figure 3.23 – Removing stopwords

While some NLP tasks (such as predicting the next word in the sentence) require stopwords, others (such as judging the sentiment of a film review) do not as the stopwords do not contribute much toward the overall meaning of the document. Removing stopwords may be preferable in these circumstances as the frequency of these common words means they can increase our feature space unnecessarily, which will increase the time it takes for our models to train.

Tagging and chunking for parts of speech

So far, we have covered several approaches for representing words and sentences, including bag-of-words, embeddings, and n-grams. However, these representations fail to capture the structure of any given sentence. Within natural language, different words can have different functions within a sentence. Consider the following:

The big dog is sleeping on the bed

We can "tag" the various words of this text, depending on the function of each word in the sentence. So, the preceding sentence becomes as follows:

The -> big -> dog -> is -> sleeping -> on -> the -> bed

Determiner -> Adjective -> Noun -> Verb -> Verb -> Preposition -> Determiner-> Noun

These **parts of speech** include, but are not limited to, the following:

Noun	Cat
Verb	Sit, stand, jumping, running
Adjective	Small, angry, sweet, clever
Adverb	Lazily, carefully, slowly, efficiently
Preposition	On, above, under, in
Determiner	The, my, his, your

Figure 3.24 – Parts of speech

These different parts of speech can be used to better understand the structure of sentences. For example, adjectives often precede nouns in English. We can use these parts of speech and their relationships to one another in our models. For example, if we are predicting the next word in the sentence and the context word is an adjective, we know the probability of the next word being a noun is high.

Tagging

Part of speech **tagging** is the act of assigning these part of speech tags to the various words within the sentence. Fortunately, NTLK has a built-in tagging functionality, so there is no need for us to train our own classifier to be able to do so:

```
sentence = "The big dog is sleeping on the bed"
token = nltk.word_tokenize(sentence)
nltk.pos_tag(token)
```

This results in the following output:

```
[('The', 'DT'),
 ('big', 'JJ'),
 ('dog', 'NN'),
 ('is', 'VBZ'),
 ('sleeping', 'VBG'),
 ('on', 'IN'),
 ('the', 'DT'),
 ('bed', 'NN')]
```

Figure 3.25 – Classifying parts of speech

Here, we simply tokenize our text and call the `pos_tag()` function to tag each of the words in the sentence. This returns a tag for each of the words in the sentence. We can decode the meaning of this tag by calling `upenn_tagset()` on the code. In this case, we can see that "VBG" corresponds to a verb:

```
nltk.help.upenn_tagset("VBG")
```

This results in the following output:

```
VBG: verb, present participle or gerund
    telegraphing stirring focusing angering judging stalling lactating
    hankerin' alleging veering capping approaching traveling besieging
    encrypting interrupting erasing wincing ...
```

Figure 3.26 – Explanation of VBG

Using pre-trained part of speech taggers is beneficial because they don't just act as a dictionary that looks up the individual words in the sentence; they also use the context of the word within the sentence to allocate its meaning. Consider the following sentences:

He drinks the water

I will buy us some drinks

The word *drinks* within these sentences represents two different parts of speech. In the first sentence, *drinks* refers to the verb; the present tense of the verb *to drink*. In the second sentence, *drinks* refers to the noun; the plural of a singular *drink*. Our pre-trained tagger is able to determine the context of these individual words and perform accurate part of speech tagging.

Chunking

Chunking expands upon our initial parts of speech tagging and aims to structure our sentences in small chunks, where each of these chunks represent a small part of speech.

We may wish to split our text up into **entities**, where each entity is a separate object or thing. For example, *the red book* refers not to three separate entities, but to a single entity described by three words. We can easily implement chunking using NLTK again. We must first define a grammar pattern to match using regular expressions. The pattern in question looks for **noun phrases (NP)**, where a noun phrase is defined as a **determiner (DT)**, followed by an **optional adjective (JJ)**, followed by a **noun (NN)**:

```
expression = ('NP: {<DT>?<JJ>*<NN>}')
```

Using the `RegexpParser()` function, we can match occurrences of this expression and tag them as noun phrases. We are then able to print the resulting tree, showing the tagged phrases. In our example sentence, we can see that *the big dog* and *the bed* are tagged as two separate noun phrases. We are able to match any chunk of text that we define using the regular expression as we see fit:

```
tagged = nltk.pos_tag(token)
REchunkParser = nltk.RegexpParser(expression)
tree = REchunkParser.parse(tagged)
print(tree)
```

This results in the following output:

Figure 3.27 – Tree representation

TF-IDF

TF-IDF is yet another technique we can learn about to better represent natural language. It is often used in text mining and information retrieval to match documents based on search terms, but can also be used in combination with embeddings to better represent sentences in embedding form. Let's take the following phrase:

This is a small giraffe

Let's say we want a single embedding to represent the meaning of this sentence. One thing we could do is simply average the individual embeddings of each of the five words in this sentence:

$$\frac{1}{5} \times \begin{bmatrix} 1 \\ 5 \\ 2 \\ 4 \end{bmatrix} \rightarrow \begin{bmatrix} 7 \\ 1 \\ 7 \\ 2 \end{bmatrix} \rightarrow \begin{bmatrix} 0 \\ 4 \\ 1 \\ 2 \end{bmatrix} \rightarrow \begin{bmatrix} 0 \\ 6 \\ 2 \\ 9 \end{bmatrix} \rightarrow \begin{bmatrix} 2 \\ 4 \\ 3 \\ 0 \end{bmatrix} = \begin{bmatrix} 2 \\ 5 \\ 3 \\ 3 \end{bmatrix}$$

Figure 3.28 – Word embeddings

However, this methodology assigns equal weight to all the words in the sentence. Do you think that all the words contribute equally to the meaning of the sentence? **This** and **a** are very common words in the English language, but **giraffe** is very rarely seen. Therefore, we might want to assign more weight to the rarer words. This methodology is known as **Term Frequency – Inverse Document Frequency (TD-IDF)**. We will now demonstrate how we can calculate TF-IDF weightings for our documents.

Calculating TF-IDF

As the name suggests, TF-IDF consists of two separate parts: term frequency and inverse document frequency. Term frequency is a document-specific measure counting the frequency of a given word within the document being analyzed:

$$tf(w, d) = \frac{count\ of\ word\ w\ in\ document\ d}{words\ in\ document\ d}$$

Note that we divide this measure by the total number of words in the document as a longer document is more likely to contain any given word. If a word appears many times in a document, it will receive a higher term frequency. However, this is the opposite of what we wish our TF-IDF weighting to do as we want to give a higher weight to occurrences of rare words within our document. This is where IDF comes into play.

Document frequency measures the number of documents within the entire corpus of documents where the word is being analyzed, and inverse document frequency calculates the ratio of the total documents to the document frequency:

$$df(w) = count\ of\ w\ across\ all\ documents$$

$$idf(w) = \frac{N}{df(w)}$$

If we have a corpus of 100 documents and our word appears five times across them, we will have an inverse document frequency of 20. This means that a higher weight is given to words with lower occurrences across all documents. Now, consider a corpus of 100,000 documents. If a word appears just once, it will have an IDF of 100,000, whereas a word occurring twice would have an IDF of 50,000. These very large and volatile IDFs aren't ideal for our calculations, so we must first normalize them with logs. Note how we add 1 within our calculations to prevent division by 0 if we calculate TF-IDF for a word that doesn't appear in our corpus:

$$idf(w) = log\left(\frac{N}{df(w) + 1}\right)$$

This makes our final TF-IDF equation look as follows:

$$tfidf(w, d) = tf(w, d) * log\left(\frac{N}{df(w) + 1}\right)$$

We can now demonstrate how to implement this in Python and apply TF-IDF weightings to our embeddings.

Implementing TF-IDF

Here, we will implement TF-IDF on a dataset using the Emma corpus from NLTK datasets. This dataset consists of a selection of sentences from the book *Emma* by Jane Austen, and we wish to calculate an embedded vector representation for each of these sentences:

1. We start by importing our dataset and looping through each of the sentences, removing any punctuation and non-alphanumeric characters (such as astericks). We choose to leave stopwords in our dataset to demonstrate how TF-IDF accounts for these as these words appear in many documents and so have a very low IDF. We create a list of parsed sentences and a set of the distinct words in our corpus:

```python
emma = nltk.corpus.gutenberg.sents('austen-emma.txt')


emma_sentences = []
emma_word_set = []


for sentence in emma:
    emma_sentences.append([word.lower() for word in
        sentence if word.isalpha()])
    for word in sentence:
        if word.isalpha():
            emma_word_set.append(word.lower())


emma_word_set = set(emma_word_set)
```

2. Next, we create a function that will return our Term Frequencies for a given word in a given document. We take the length of the document to give us the number of words and count the occurrences of this word in the document before returning the ratio. Here, we can see that the word ago appears in the sentence once and that the sentence is 41 words long, giving us a Term Frequency of 0.024:

```python
def TermFreq(document, word):
    doc_length = len(document)
    occurances = len([w for w in document if w == word])
    return occurances / doc_length


TermFreq(emma_sentences[5], 'ago')
```

This results in the following output:

0.024390243902439025

Figure 3.29 – TF-IDF score

3. Next, we calculate our Document Frequency. In order to do this efficiently, we first need to pre-compute a Document Frequency dictionary. This loops through all the data and counts the number of documents each word in our corpus appears in. We pre-compute this so we that do not have to perform this loop every time we wish to calculate Document Frequency for a given word:

```
def build_DF_dict():
    output = {}
    for word in emma_word_set:
        output[word] = 0
        for doc in emma_sentences:
            if word in doc:
                output[word] += 1
    return output


df_dict = build_DF_dict()


df_dict['ago']
```

4. Here, we can see that the word ago appears within our document 32 times. Using this dictionary, we can very easily calculate our Inverse Document Frequency by dividing the total number of documents by our Document Frequency and taking the logarithm of this value. Note how we add one to the Document Frequency to avoid a divide by zero error when the word doesn't appear in the corpus:

```
def InverseDocumentFrequency(word):
    N = len(emma_sentences)
    try:
        df = df_dict[word] + 1
    except:
        df = 1
    return np.log(N/df)


InverseDocumentFrequency('ago')
```

5. Finally, we simply combine the Term Frequency and Inverse Document Frequency to get the TF-IDF weighting for each word/document pair:

```
def TFIDF(doc, word):
    tf = TF(doc, word)
    idf = InverseDocumentFrequency(word)
    return tf*idf

print('ago - ' + str(TFIDF(emma_sentences[5],'ago')))
print('indistinct - ' + str(TFIDF(emma_
sentences[5],'indistinct')))
```

This results in the following output:

```
ago - 0.13315118517327126
indistinct - 0.20152582861001603
```

Figure 3.30 – TF-IDF score for ago and indistinct

Here, we can see that although the words ago and indistinct appear only once in the given document, indistinct occurs less frequently throughout the whole corpus, meaning it receives a higher TF-IDF weighting.

Calculating TF-IDF weighted embeddings

Next, we can show how these TF-IDF weightings can be applied to embeddings:

1. We first load our pre-computed GLoVe embeddings to provide the initial embedding representation of words in our corpus:

```
def loadGlove(path):
    file = open(path,'r')
    model = {}
    for l in file:
        line = l.split()
        word = line[0]
        value = np.array([float(val) for val in
                          line[1:]])
        model[word] = value
    return model

glove = loadGlove('glove.6B.50d.txt')
```

2. We then calculate an unweighted mean average of all the individual embeddings in our document to get a vector representation of the sentence as a whole. We simply loop through all the words in our document, extract the embedding from the GLoVe dictionary, and calculate the average over all these vectors:

```
embeddings = []


for word in emma_sentences[5]:
    embeddings.append(glove[word])


mean_embedding = np.mean(embeddings, axis = 0).reshape
    (1, -1)


print(mean_embedding)
```

This results in the following output:

```
[[ 3.32575634e-01  3.16596488e-01 -1.80050732e-01 -3.82070951e-01
   4.98493527e-01  5.33804805e-01 -5.46517073e-01  9.12476195e-02
  -1.31538483e-01 -2.71967805e-02  2.99867317e-02  2.64278024e-02
  -2.06519756e-01 -1.54796634e-01  4.28036366e-01 -5.74977317e-02
  -2.65928778e-01  1.60373902e-02 -2.84913561e-01 -2.01252268e-01
  -5.96390732e-02  5.72458220e-01  2.06195927e-01 -1.54312293e-01
   2.52049805e-01 -1.64638200e+00 -3.42686049e-01  1.02592522e-01
   1.42848000e-01 -1.09779902e-01  2.89345488e+00  7.36985634e-02
  -3.73648780e-03 -2.76292784e-01  1.50580049e-01  9.80399951e-02
   2.24408780e-03  2.83664024e-01  3.92979024e-02 -2.98091634e-01
  -1.17309171e-01  2.08815776e-01  6.89953902e-03  2.92777244e-02
   5.54180122e-02 -2.20519707e-01 -2.82007805e-01 -4.34917439e-01
  -9.69051537e-02 -1.67569878e-01]]
```

Figure 3.31 – Mean embedding

3. We repeat this process to calculate our TF-IDF weighted document vector, but this time, we multiply our vectors by their TF-IDF weighting before we average them:

```
embeddings = []


for word in emma_sentences[5]:
    tfidf = TFIDF(emma_sentences[5], word)
    embeddings.append(glove[word] * tfidf)


tfidf_weighted_embedding = np.mean(embeddings, axis =
                            0).reshape(1, -1)


print(tfidf_weighted_embedding)
```

This results in the following output:

```
[[ 0.03390627   0.04567951 -0.02513047 -0.05553374   0.06523389   0.07031937
  -0.06309126   0.02674499 -0.01073998 -0.00509068   0.00518551   0.00818713
  -0.01610237 -0.01486281   0.04954961 -0.0107796  -0.05029558   0.00039276
  -0.0192399  -0.01344365 -0.01123742   0.08506534   0.02145731 -0.0159164
   0.04411737 -0.17889813 -0.04006272   0.01603446   0.02090289 -0.01344211
   0.28346797   0.00696015   0.00484046 -0.02637939   0.01537125   0.01611019
   0.00316879   0.0324516    0.00829024 -0.04200008 -0.0058922    0.01996137
  -0.00305491 -0.00355021   0.01175475 -0.03423196 -0.02943769 -0.06810232
  -0.00775695 -0.0181068 ]]
```

Figure 3.32 – TF-IDF embedding

4. We can then compare the TF-IDF weighted embedding with our average embedding to see how similar they are. We can do this using cosine similarity, as follows:

```
cosine_similarity(mean_embedding, tfidf_weighted_
embedding)
```

This results in the following output:

```
array([[0.98653879]])
```

Figure 3.33 – Cosine similarity between TF-IDF and average embedding

Here, we can see that our two different representations are very similar. Therefore, while using TF-IDF may not dramatically change our representation of a given sentence or document, it may weigh it in favor of words of interest, thus providing a more useful representation.

Summary

In this chapter, we have taken a deeper dive into word embeddings and their applications. We have demonstrated how they can be trained using a continuous bag-of-words model and how we can incorporate n-gram language modeling to better understand the relationship between words in a sentence. We then looked at splitting documents into individual tokens for easy processing and how to use tagging and chunking to identify parts of speech. Finally, we showed how TF-IDF weightings can be used to better represent documents in embedding form.

In the next chapter, we will see how to use NLP for text preprocessing, stemming, and lemmatization.

4
Text Preprocessing, Stemming, and Lemmatization

Textual data can be gathered from a number of different sources and takes many different forms. Text can be tidy and readable or raw and messy and can also come in many different styles and formats. Being able to preprocess this data so that it can be converted into a standard format before it reaches our NLP models is what we'll be looking at in this chapter.

Stemming and lemmatization, similar to tokenization, are other forms of NLP preprocessing. However, unlike tokenization, which reduces a document into individual words, stemming and lemmatization are attempts to reduce these words further to their lexical roots. For example, almost any verb in English has many different variations, depending on tense:

He jumped

He is jumping

He jumps

While all these words are different, they all relate to the same root word – **jump**. Stemming and lemmatization are both techniques we can use to reduce word variations to their common roots.

In this chapter, we will explain how to perform preprocessing on textual data, as well as explore both stemming and lemmatization and show how these can be implemented in Python.

In this chapter, we will cover the following topics:

- Text preprocessing
- Stemming
- Lemmatization
- Uses of stemming and lemmatization

Technical requirements

For the text preprocessing in this chapter, we will mostly use inbuilt Python functions, but we will also use the external `BeautifulSoup` package. For stemming and lemmatization, we will use the NLTK Python package. All the code in this chapter can be found at `https://github.com/PacktPublishing/Hands-On-Natural-Language-Processing-with-PyTorch-1.x/tree/master/Chapter4`.

Text preprocessing

Textual data can come in a variety of formats and styles. Text may be in a structured, readable format or in a more raw, unstructured format. Our text may contain punctuation and symbols that we don't wish to include in our models or may contain HTML and other non-textual formatting. This is of particular concern when scraping text from online sources. In order to prepare our text so that it can be input into any NLP models, we must perform preprocessing. This will clean our data so that it is in a standard format. In this section, we will illustrate some of these preprocessing steps in more detail.

Removing HTML

When scraping text from online sources, you may find that your text contains HTML markup and other non-textual artifacts. We do not generally want to include these in our NLP inputs for our models, so these should be removed by default. For example, in HTML, the <b> tag indicates that the text following it should be in bold font. However, this does not contain any textual information about the content of the sentence, so we should remove this. Fortunately, in Python, there is a package called BeautifulSoup that allows us to remove all HTML in a few lines:

```
input_text = "<b> This text is in bold</br>, <i> This text is
in italics </i>"
output_text =  BeautifulSoup(input_text, "html.parser").get_
text()
print('Input: ' + input_text)
print('Output: ' + output_text)
```

This returns the following output:

```
Input: <b> This text is in bold</br>, <i> This text is in italics </i>
Output:  This text is in bold,  This text is in italics
```

Figure 4.1 – Removing HTML

The preceding screenshot shows that the HTML has been successfully removed. This could be useful in any situations where HTML code may be present within raw text data, such as when scraping a web page for data.

Converting text into lowercase

It is standard practice when preprocessing text to convert everything into lowercase. This is because any two words that are the same should be considered semantically identical, regardless of whether they are capitalized or not. 'Cat', 'cat', and 'CAT' are all the same words but just have different elements capitalized. Our models will generally consider these three words as separate entities as they are not identical. Therefore, it is standard practice to convert all words into lowercase so that these words are all semantically and structurally identical. This can be done very easily within Python using the following lines of code:

```
input_text = ['Cat','cat','CAT']
output_text =  [x.lower() for x in input_text]
print('Input: ' + str(input_text))
print('Output: ' + str(output_text))
```

This returns the following output:

```
Input: ['Cat', 'cat', 'CAT']
Output: ['cat', 'cat', 'cat']
```

Figure 4.2 – Converting input into lowercase

This shows that the inputs have all been transformed into identical lowercase representations. There are a few examples where capitalization may actually provide additional semantic information. For example, *May* (the month) and *may* (meaning *might*) are semantically different and *May* (the month) will always be capitalized. However, instances like this are very rare and it is much more efficient to convert everything into lowercase than trying to account for these rare examples.

It is worth noting that capitalization may be useful in some tasks such as part of speech tagging, where a capital letter may indicate the word's role in the sentence, and named entity recognition, where a capital letter may indicate that a word is a proper noun rather than the non-proper noun alternative; for example, *Turkey* (the country) and *turkey* (the bird).

Removing punctuation

Sometimes, depending on the type of model being constructed, we may wish to remove punctuation from our input text. This is particularly useful in models where we are aggregating word counts, such as in a bag-of-words representation. The presence of a full stop or a comma within the sentence doesn't add any useful information about the semantic content of the sentence. However, more complicated models that take into account the position of punctuation within the sentence may actually use the position of the punctuation to infer a different meaning. A classic example is as follows:

The panda eats shoots and leaves

The panda eats, shoots, and leaves

Here, the addition of a comma transforms the sentence describing a panda's eating habits into a sentence describing an armed robbery of a restaurant by a panda! Nevertheless, it is still important to be able to remove punctuation from sentences for the sake of consistency. We can do this in Python by using the re library, to match any punctuation using a regular expression, and the sub() method, to replace any matched punctuation with an empty character:

```
input_text = "This ,sentence.'' contains-£ no:: punctuation?"
output_text = re.sub(r'[^\w\s]', '', input_text)
print('Input: ' + input_text)
print('Output: ' + output_text)
```

This returns the following output:

```
Input: This ,sentence.'' contains-£ no:: puncuation?
Output: This sentence contains no puncuation
```

Figure 4.3 – Removing punctuation from input

This shows that the punctuation has been removed from the input sentence.

There may be instances where we may not wish to directly remove punctuation. A good example would be the use of the ampersand (&), which in almost every instance is used interchangeably with the word "and". Therefore, rather than completely removing the ampersand, we may instead opt to replace it directly with the word "and". We can easily implement this in Python using the .replace() function:

```
input_text = "Cats & dogs"
output_text = input_text.replace("&", "and")
print('Input: ' + input_text)
print('Output: ' + output_text)
```

This returns the following output:

```
Input: Cats & dogs
Output: Cats and dogs
```

Figure 4.4 – Removing and replacing punctuation

It is also worth considering specific circumstances where punctuation may be essential for the representation of a sentence. One crucial example is email addresses. Removing the @ from email addresses doesn't make the address any more readable:

```
name@gmail.com
```

Removing the punctuation returns this:

namegmailcom

So, in instances like this, it may be preferable to remove the whole item altogether, according to the requirements and purpose of your NLP model.

Replacing numbers

Similarly, with numbers, we also want to standardize our outputs. Numbers can be written as digits (9, 8, 7) or as actual words (nine, eight, seven). It may be worth transforming these all into a single, standardized representation so that 1 and one are not treated as separate entities. We can do this in Python using the following methodology:

```python
def to_digit(digit):
    i = inflect.engine()
    if digit.isdigit():
        output = i.number_to_words(digit)
    else:
        output = digit
    return output

input_text = ["1","two","3"]
output_text = [to_digit(x) for x in input_text]
print('Input: ' + str(input_text))
print('Output: ' + str(output_text))
```

This returns the following output:

```
Input: ['1', 'two', '3']
Output: ['one', 'two', 'three']
```

Figure 4.5 – Replacing numbers with text

This shows that we have successfully converted our digits into text.

However, in a similar fashion to processing email addresses, processing phone numbers may not require the same representation as regular numbers. This is illustrated in the following example:

```python
input_text = ["0800118118"]
output_text = [to_digit(x) for x in input_text]
print('Input: ' + str(input_text))
print('Output: ' + str(output_text))
```

This returns the following output:

```
Input: ['0800118118']
Output: ['eight hundred million, one hundred and eighteen thousand, one hundred and eighteen']
```

Figure 4.6 – Converting a phone number into text

Clearly, the input in the preceding example is a phone number, so the full text representation is not necessarily fit for purpose. In instances like this, it may be preferable to drop any long numbers from our input text.

Stemming and lemmatization

In language, **inflection** is how different grammatical categories such as tense, mood, or gender can be expressed by modifying a common root word. This often involves changing the prefix or suffix of a word but can also involve modifying the entire word. For example, we can make modifications to a verb to change its tense:

Run -> Runs (Add "s" suffix to make it present tense)

Run -> Ran (Modify middle letter to "a" to make it past tense)

But in some cases, the whole word changes:

To be -> Is (Present tense)

To be -> Was (Past tense)

To be -> Will be (Future tense – addition of modal)

There can be lexical variations on nouns too:

Cat -> Cats (Plural)

Cat -> Cat's (Possessive)

Cat -> Cats' (Plural possessive)

All these words relate back to the root word cat. We can calculate the root of all the words in the sentence to reduce the whole sentence to its lexical roots:

"His cats' fur are different colors" -> "He cat fur be different color"

Stemming and lemmatization is the process by which we arrive at these root words. **Stemming** is an algorithmic process in which the ends of words are cut off to arrive at a common root, whereas lemmatization uses a true vocabulary and structural analysis of the word itself to arrive at the true roots, or **lemmas**, of the word. We will cover both of these methodologies in detail in the following sections.

Stemming

Stemming is the algorithmic process by which we trim the ends off words in order to arrive at their lexical roots, or **stems**. To do this, we can use different **stemmers** that each follow a particular algorithm in order to return the stem of a word. In English, one of the most common stemmers is the Porter Stemmer.

The **Porter Stemmer** is an algorithm with a large number of logical rules that can be used to return the stem of a word. We will first show how to implement a Porter Stemmer in Python using NLTK before moving on and discussing the algorithm in more detail:

1. First, we create an instance of the Porter Stemmer:

    ```
    porter = PorterStemmer()
    ```

2. We then simply call this instance of the stemmer on individual words and print the results. Here, we can see an example of the stems returned by the Porter Stemmer:

    ```
    word_list = ["see","saw","cat", "cats", "stem",
    "stemming","lemma","lemmatization","known","knowing","time",
    "timing","football", "footballers"]
    for word in word_list:
        print(word + ' -> ' + porter.stem(word))
    ```

    This results in the following output:

    ```
    see -> see
    saw -> saw
    cat -> cat
    cats -> cat
    stem -> stem
    stemming -> stem
    lemma -> lemma
    lemmatization -> lemmat
    known -> known
    knowing -> know
    time -> time
    timing -> time
    football -> footbal
    footballers -> footbal
    ```

    Figure 4.7 – Returning the stems of words

3. We can also apply stemming to an entire sentence, first by tokenizing the sentence and then by stemming each term individually:

```
def SentenceStemmer(sentence):
    tokens=word_tokenize(sentence)
    stems=[porter.stem(word) for word in tokens]
    return " ".join(stems)
SentenceStemmer('The cats and dogs are running')
```

This returns the following output:

```
'the cat and dog are run'
```

Figure 4.8 – Applying stemming to a sentence

Here, we can see how different words are stemmed using the Porter Stemmer. Some words, such as stemming and timing, reduce to their expected stems of stem and time. However, some words, such as saw, don't reduce to their logical stem (see). This illustrates the limitations of the Porter Stemmer. Since stemming applies a series of logical rules to the word, it is very difficult to define a set of rules that will correctly stem all words. This is especially true in the cases of words in English where the word changes completely, depending on the tense (is/was/be). This is because there are no generic rules that can be applied to these words to transform them all into the same root stem.

We can examine some of the rules the Porter Stemmer applies in more detail to understand exactly how the transformation into the stem occurs. While the actual Porter algorithm has many detailed steps, here, we will simplify some of the rules for ease of understanding:

Rule	Example
"sses" becomes "ss"	Dresses -> Dress
"s" becomes " "	Dogs -> Dog
"eed" becomes "ee" for longer words	Agreed -> Agree Seed -> Seed
"ed" becomes " " if the word contains vowels before the "ed"	Flexed -> Flex Bed -> Bed
"ing" becomes " " if the word contains vowels before the "ing"	Jumping -> Jump Ring -> Ring

Figure 4.9 – Rules of the Porter Stemmer algorithm

While it is not essential to understand every rule within the Porter Stemmer, it is key that we understand its limitations. While the Porter Stemmer has been shown to work well across a corpus, there will always be words that it cannot reduce to their true stems correctly. Since the rule set of the Porter Stemmer relies on the conventions of English word structure, there will always be words that do not fall within the conventional word structure and are not correctly transformed by these rules. Fortunately, some of these limitations can be overcome through the use of lemmatization.

Lemmatization

Lemmatization differs from stemming in that it reduces words to their **lemma** instead of their stem. While the stem of a word is processed and reduced to a string, a word's lemma is its true lexical root. So, while the stem of the word `ran` will just be *ran*, its lemma is the true lexical root of the word, which would be `run`.

The lemmatization process uses both inbuilt pre-computed lemmas and associated words, as well as the context of the word within the sentence to determine the correct lemma for a given word. In this example, we will look at using the **WordNet Lemmatizer** within NLTK. WordNet is a large database of English words and their lexical relationships to one another. It contains one of the most robust and comprehensive mappings of the English language, specifically with regard to words' relationships to their lemmas.

We will first create an instance of our lemmatizer and call it on a selection of words:

```
wordnet_lemmatizer = WordNetLemmatizer()

print(wordnet_lemmatizer.lemmatize('horses'))
print(wordnet_lemmatizer.lemmatize('wolves'))
print(wordnet_lemmatizer.lemmatize('mice'))
print(wordnet_lemmatizer.lemmatize('cacti'))
```

This results in the following output:

```
horse
wolf
mouse
cactus
```

Figure 4.10 – Lemmatization output

Here, we can already begin to see the advantages of using lemmatization over stemming. Since the WordNet Lemmatizer is built on a database of all the words in the English language, it knows that `mice` is the plural version of `mouse`. We would not have been able to reach this same root using stemming. Although lemmatization works better in the majority of cases, because it relies on a built-in index of words, it is not able to generalize to new or made-up words:

```
print(wordnet_lemmatizer.lemmatize('madeupwords'))
print(porter.stem('madeupwords'))
```

This results in the following output:

```
madeupwords
madeupword
```

Figure 4.11 – Lemmatization output for made-up words

Here, we can see that, in this instance, our stemmer is able to generalize better to previously unseen words. Therefore, using a lemmatizer may be a problem if we're lemmatizing sources where language doesn't necessarily match up with *real* English language, such as social media sites where people may frequently abbreviate language.

If we call our lemmatizer on two verbs, we will see that this doesn't reduce them to their expected common lemma:

```
print(wordnet_lemmatizer.lemmatize('run'))
print(wordnet_lemmatizer.lemmatize('ran'))
```

This results in the following output:

```
ran
run
```

Figure 4.12 – Running lemmatization on verbs

This is because our lemmatizer relies on the context of words to be able to return the lemmas. Recall from our POS analysis that we can easily return the context of a word in a sentence and determine whether a given word is a noun, verb, or adjective. For now, let's manually specify that our words are verbs. We can see that this now correctly returns the lemma:

```
print(wordnet_lemmatizer.lemmatize('ran', pos='v'))
print(wordnet_lemmatizer.lemmatize('run', pos='v'))
```

This results in the following output:

```
run
run
```

Figure 4.13 – Implementing POS in the function

This means that in order to return the correct lemmatization of any given sentence, we must first perform POS tagging to obtain the context of the words in the sentence, then pass this through the lemmatizer to obtain the lemmas of each of the words in the sentence. We first create a function that will return our POS tagging for each word in the sentence:

```
sentence = 'The cats and dogs are running'

def return_word_pos_tuples(sentence):
    return nltk.pos_tag(nltk.word_tokenize(sentence))

return_word_pos_tuples(sentence)
```

This results in the following output:

```
[('The', 'DT'),
 ('cats', 'NNS'),
 ('and', 'CC'),
 ('dogs', 'NNS'),
 ('are', 'VBP'),
 ('running', 'VBG')]
```

Figure 4.14 – Output of POS tagging on a sentence

Note how this returns the NLTK POS tags for each of the words in the sentence. Our WordNet lemmatizer requires a slightly different input for POS. This means that we first create a function that maps the NLTK POS tags to the required WordNet POS tags:

```
def get_pos_wordnet(pos_tag):
    pos_dict = {"N": wordnet.NOUN,
                "V": wordnet.VERB,
                "J": wordnet.ADJ,
                "R": wordnet.ADV}

    return pos_dict.get(pos_tag[0].upper(), wordnet.NOUN)

get_pos_wordnet('VBG')
```

This results in the following output:

$$'v'$$

Figure 4.15 – Mapping NTLK POS tags to WordNet POS tags

Finally, we combine these functions into one final function that will perform lemmatization on the whole sentence:

```
def lemmatize_with_pos(sentence):
    new_sentence = []
    tuples = return_word_pos_tuples(sentence)
    for tup in tuples:
        pos = get_pos_wordnet(tup[1])
        lemma = wordnet_lemmatizer.lemmatize(tup[0], pos=pos)
        new_sentence.append(lemma)
    return new_sentence

lemmatize_with_pos(sentence)
```

This results in the following output:

```
['The', 'cat', 'and', 'dog', 'be', 'run']
```

Figure 4.16 – Output of the finalized lemmatization function

Here, we can see that, in general, lemmas generally provide a better representation of a word's true root compared to stems, with some notable exceptions. When we might decide to use stemming and lemmatization depends on the requirements of the task at hand, some of which we will explore now.

Uses of stemming and lemmatization

Stemming and lemmatization are both a form of NLP that can be used to extract information from text. This is known as **text mining**. Text mining tasks come in a variety of categories, including text clustering, categorization, summarizing documents, and sentiment analysis. Stemming and lemmatization can be used in conjunction with deep learning to solve some of these tasks, as we will see later in this book.

By performing preprocessing using stemming and lemmatization, coupled with the removal of stop words, we can better reduce our sentences to understand their core meaning. By removing words that do not significantly contribute to the meaning of the sentence and by reducing words to their roots or lemmas, we can efficiently analyze sentences within our deep learning frameworks. If we are able to reduce a 10-word sentence to five words consisting of multiple core lemmas rather than multiple variations of similar words, this means much less data that we need to feed through our neural networks. If we use bag-of-words representations, our corpus will be significantly smaller as multiple words all reduce down to the same lemmas, whereas if we calculate embedding representations, the dimensionality required to capture the true representations of our words will be smaller for a reduced corpus of words.

Differences in lemmatization and stemming

Now that we have seen both lemmatization and stemming in action, the question still remains as to under which circumstances we should use both of these techniques. We saw that both techniques attempt to reduce each word to its root. In stemming, this may just be a reduced form of the target room, whereas in lemmatization, it reduces to a true English language word root.

Because lemmatization requires cross-referencing the target word within the WordNet corpus, as well as performing part-of-speech analysis to determine the form of the lemma, this may take a significant amount of processing time if a large number of words have to be lemmatized. This is in contrast to stemming, which uses a detailed but relatively fast algorithm to stem words. Ultimately, as with many problems in computing, it is a question of trading off speed versus detail. When choosing which of these methods to incorporate in our deep learning pipeline, the trade-off may be between speed and accuracy. If time is of the essence, then stemming may be the way to go. On the other hand, if you need your model to be as detailed and as accurate as possible, then lemmatization will likely result in the superior model.

Summary

In this chapter, we have covered both stemming and lemmatization in detail by exploring the functionality of both methods, their use cases, and how they can be implemented. Now that we have covered all of the fundamentals of deep learning and NLP preprocessing, we are ready to start training our own deep learning models from scratch.

In the next chapter, we will explore the fundamentals of NLP and demonstrate how to build the most widely used models within the field of deep NLP: recurrent neural networks.

Section 3: Real-World NLP Applications Using PyTorch 1.x

In this section, we will use the various **Natural Language Processing (NLP)** techniques available in PyTorch to build a wide range of real-world applications using PyTorch. Sentiment analysis, text summarization, text classification, and building a chatbot application using PyTorch are some of the tasks that will be covered in this section.

This section contains the following chapters:

- *Chapter 5, Recurrent Neural Networks and Sentiment Analysis*
- *Chapter 6, Convolutional Neural Networks for Text Classification*
- *Chapter 7, Text Translation Using Sequence-to-Sequence Neural Networks*
- *Chapter 8, Building a Chatbot Using Attention-Based Neural Networks*
- *Chapter 9, The Road Ahead*

5
Recurrent Neural Networks and Sentiment Analysis

In this chapter, we will look at **Recurrent Neural Networks** (**RNNs**), a variation of the basic feed forward neural networks in PyTorch that we learned how to build in *Chapter 1, Fundamentals of Machine Learning*. Generally, RNNs can be used for any task where data can be represented as a sequence. This includes things such as stock price prediction, using a time series of historic data represented as a sequence. We commonly use RNNs in NLP as text can be thought of as a sequence of individual words and can be modeled as such. While a conventional neural network takes a single vector as input to the model, an RNN can take a whole sequence of vectors. If we represent each word in a document as a vector embedding, we can represent a whole document as a sequence of vectors (or an order 3 tensor). We can then use RNNs (and a more sophisticated form of RNN known as **Long Short-Term Memory** (**LSTM**) to learn from our data.

In this chapter, we will cover the basics of RNNs and the more advanced LSTM. We will then look at sentiment analysis and work through a practical example of how to build an LSTM to classify documents using PyTorch. Finally, we will host our simple model on Heroku, a simple cloud application platform, which will allow us to make predictions using our model.

This chapter covers the following topics:

- Building RNNs
- Working with LSTMs
- Building a sentiment analyzer using LSTM
- Deploying the application on Heroku

Technical requirements

All the code used in this chapter can be found at `https://github.com/PacktPublishing/Hands-On-Natural-Language-Processing-with-PyTorch-1.x/tree/master/Chapter5`. Heroku can be installed from `www.heroku.com`. The data was taken from `https://archive.ics.uci.edu/ml/datasets/Sentiment+Labelled+Sentences`.

Building RNNs

RNNs consist of recurrent layers. While they are similar in many ways to the fully connected layers within a standard feed forward neural network, these recurrent layers consist of a hidden state that is updated at each step of the sequential input. This means that for any given sequence, the model is initialized with a hidden state, often represented as a one-dimensional vector. The first step of our sequence is then fed into our model and the hidden state is updated depending on some learned parameters. The second word is then fed into the network and the hidden state is updated again depending on some other learned parameters. These steps are repeated until the whole sequence has been processed and we are left with the final hidden state. This computation *loop*, with the hidden state carried over from the previous computation and updated, is why we refer to these networks as recurrent. This final hidden state is then connected to a further fully connected layer and a final classification is predicted.

Our recurrent layer looks something like the following, where h is the hidden state and x is our input at various time steps in our sequence. For each iteration, we update our hidden state at each time step, x:

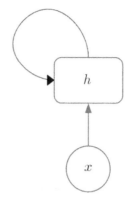

Figure 5.1 – Recurrent layer

Alternatively, we can expand this out to the whole sequence of time steps, which looks like this:

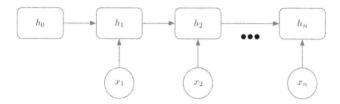

Figure 5.2 – Sequence of time steps

This layer is for an input that is n time steps long. Our hidden state is initialized in state h_0, and then uses our first input, x_1, to compute the next hidden state, h_1. There are two sets of weight matrices that are also learned—matrix U, which learns how the hidden state changes between time steps, and matrix W, which learns how each input step affects the hidden state.

We also apply a *tanh* activation function to the resulting product, keeping the values of the hidden state between -1 and 1. The equation for calculating any hidden state, h_t, becomes the following:

$$h_t = \tanh (W h_{t-1} + U x_t)$$

This is then repeated for each time step within our input sequence, and the final output for this layer is our last hidden state, h_n. When our network learns, we perform a forward pass through the network, as before, to compute our final classification. We then calculate a loss against this prediction and backpropagate through the network, as before, calculating gradients as we go. This backpropagation process occurs through all the steps within the recurrent layer, with the parameters between each input step and the hidden state being learned.

We will see later that we can actually take the hidden state at each time step, rather than using the final hidden state, which is useful for sequence-to-sequence translation tasks in NLP. However, for the time being, we will just take the hidden layer as output to the rest of the network.

Using RNNs for sentiment analysis

In the context of sentiment analysis, our model is trained on a sentiment analysis dataset of reviews that consists of a number of reviews in text and a label of 0 or 1, depending on whether the review is negative or positive. This means that our model becomes a classification task (where the two classes are negative/positive). Our sentence is passed through a layer of learned word embeddings to form a representation of the sentence comprising several vectors (one for each word). These vectors are then fed sequentially into our RNN layer and the final hidden state is passed through another fully connected layer. Our model's output is a single value between 0 and 1, depending on whether our model predicts a negative or positive sentiment from the sentence. This means our complete classification model looks like this:

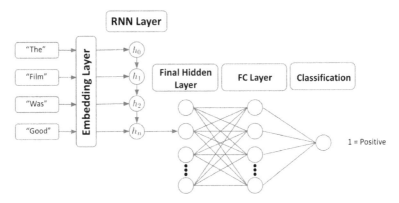

Figure 5.3 – Classification model

Now, we will highlight one of the issues with RNNs—exploding and shrinking gradients—and how we can remedy this using gradient clipping.

Exploding and shrinking gradients

One issue that we are often faced with within RNNs is that of **exploding or shrinking gradients**. We can think of the recursive layer as a very deep network. When calculating the gradients, we do so at every iteration of the hidden state. If the gradient of the loss relative to the weights at any given position becomes very big, this will have a multiplicative effect as it feeds forward through all the iterations of the recurrent layer. This can cause gradients to explode as they get very large very quickly. If we have large gradients, this can cause instability in our network. On the other hand, if the gradients within our hidden state are very small, this will again have a multiplicative effect and the gradients will be close to 0. This means that the gradients can become too small to accurately update our parameters via gradient descent, meaning our model fails to learn.

One technique we can use to prevent our gradients from exploding is to use **gradient clipping**. This technique limits our gradients to prevent them from becoming too large. We simply choose a hyperparameter, C, and can calculate our clipped gradient, as follows:

$$\left\lVert \lVert \nabla L \rVert \right\rVert_c = \max \left(\lVert \nabla L \rVert, C \right)$$

The following graph shows the relationship between the two variables:

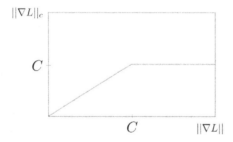

Figure 5.4 – Comparison of gradient clipping

Another technique we can use to prevent exploding or disappearing gradients is to shorten our input sequence length. The effective depth of our recurrent layer depends on the length of our input sequence as the sequence length determines how many iterative updates we need to perform on our hidden state. The fewer number of steps in this process, the smaller the multiplicative effects of the gradient accumulation between hidden states will be. By intelligently picking the maximum sequence length as a hyperparameter in our model, we can help prevent exploding and vanishing gradients.

Introducing LSTMs

While RNNs allow us to use sequences of words as input to our models, they are far from perfect. RNNs suffer from two main flaws, which can be partially remedied by using a more sophisticated version of the RNN, known as **LSTM**.

The basic structure of RNNs means that it is very difficult for them to retain information long term. Consider a sentence that's 20 words long. From our first word in the sentence affecting the initial hidden state to the last word in the sentence, our hidden state is updated 20 times. From the beginning of our sentence to our final hidden state, it is very difficult for an RNN to retain information about words at the beginning of the sentence. This means that RNNs aren't very good at capturing long-term dependencies within sequences. This also ties in with the vanishing gradient problem mentioned earlier, where it is very inefficient to backpropagate through long, sparse sequences of vectors.

Consider a long paragraph where we are trying to predict the next word. The sentence begins with I study math... and ends with my final exam is in.... Intuitively, we would expect the next word to be math or some math-related field. However, in an RNN model on a long sequence, our hidden state may struggle to retain the information for the beginning of the sentence by the time it reaches the end of the sentence as it takes multiple update steps.

We should also note that RNNs are poor at capturing the context of words within a sentence as a whole. We saw earlier, when looking at n-gram models, that the meaning of a word in a sentence is dependent on its context within the sentence, which is determined by the words that occur before it and the words that occur after it. Within an RNN, our hidden state updates in one direction only. In a single forward pass, our hidden state is initialized and the first word in the sequence is passed into it. This process is then repeated with all the subsequent words in the sentence sequentially until we are left with our final hidden state. This means that for any given word in a sentence, we have only considered the cumulative effect of the words that have occurred before it in the sentence up to that point. We do not consider any words that follow it, meaning we do not capture the full context of each word in the sentence.

In another example, we again want to predict the missing word in a sentence, but it now occurs toward the beginning as opposed to at the end. We have the sentence I grew up in...so I can speak fluent Dutch. Here, we can intuitively guess that the person grew up in the Netherlands from the fact that they speak Dutch. However, because an RNN parses this information sequentially, it would only use I grew up in... to make a prediction, missing the other key context within the sentence.

Both of these issues can be partially addressed using LSTMs.

Working with LSTMs

LSTMs are more advanced versions of RNNs and contain two extra properties—an **update gate** and a **forget gate**. These two additions make it easier for the network to learn long-term dependencies. Consider the following film review:

The film was amazing. I went to see it with my wife and my daughters on Tuesday afternoon. Although I didn't expect it to be very entertaining, it turned out to be loads of fun. We would definitely go back and see it again given the chance.

In sentiment analysis, it is clear that not all of the words in the sentence are relevant in determining whether it is a positive or negative review. We will repeat this sentence, but this time highlighting the words that are relevant to gauging the sentiment of the review:

The film was amazing. I went to see it with my wife and my daughters on Tuesday afternoon. Although I didn't expect it to be very entertaining, it turned out to be loads of fun. We would definitely go back and see it again given the chance.

LSTMs attempt to do exactly this—remember the relevant words within a sentence while forgetting all the irrelevant information. By doing this, it stops the irrelevant information from diluting the relevant information, meaning long-term dependencies can be better learned across long sequences.

LSTMs are very similar in structure to RNNs. While there is a hidden state that is carried over between steps within the LSTM, the inner workings of the LSTM cell itself are different from that of the RNN:

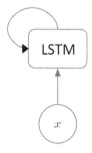

Figure 5.5 – LSTM cell

LSTM cells

While an RNN cell just takes the previous hidden state and the new input step and calculates the next hidden state using some learned parameters, the inner workings of an LSTM cell are significantly more complicated:

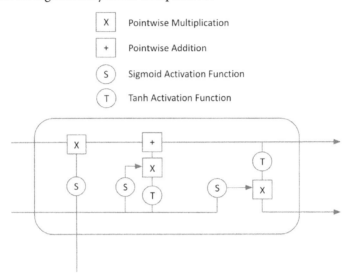

Figure 5.6 – Inner workings of an LSTM cell

While this looks significantly more daunting than the RNN, we will explain each component of the LSTM cell in turn. We will first look at the **forget gate** (indicated by the bold rectangle):

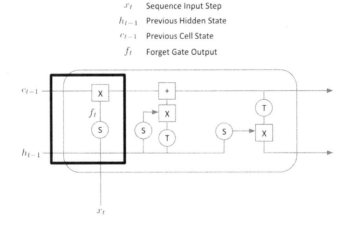

Figure 5.7 – The forget gate

The forget gate essentially learns which elements of the sequence to forget. The previous hidden state, h_{t-1}, and the latest input step, x_1, are concatenated together and passed through a matrix of learned weights on the forget gate and a sigmoid function that squashes the values between 0 and 1. This resulting matrix, ft, is multiplied pointwise by the cell state from the previous step, c_{t-1}. This effectively applies a mask to the previous cell state so that only the relevant information from the previous cell state is brought forward.

Next, we will look at the **input gate**:

x_t	Sequence Input Step
h_{t-1}	Previous Hidden State
c_{t-1}	Previous Cell State
i_t	Input Gate Output

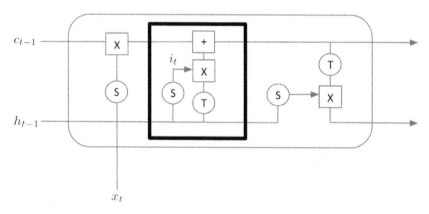

Figure 5.8 – The input gate

The input gate again takes the concatenated previous hidden state, h_{t-1}, and the current sequence input, x_t, and passes this through a sigmoid function with learned parameters, which outputs another matrix, i_t, consisting of values between 0 and 1. The concatenated hidden state and sequence input also pass through a tanh function, which squashes the output between -1 and 1. This is multiplied by the i_t matrix. This means that the learned parameters required to generate i_t effectively learn which elements should be kept from the current time step in our cell state. This is then added to the current cell state to get our final cell state, which will be carried over to the next time step.

Finally, we have the last element of the LSTM cell—the **output gate**:

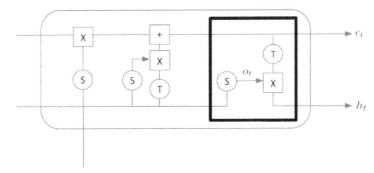

Figure 5.9 – The output gate

The output gate calculates the final output of the LSTM cell—both the cell state and the hidden state that is carried over to the next step. The cell state, c_t, is unchanged from the previous two steps and is a product of the forget gate and the input gate. The final hidden state, h_t, is calculated by taking the concatenated previous hidden state, h_{t-1}, and the current time step input, x_t, and passing through a sigmoid function with some learned parameters to get the output gate output, o_t. The final cell state, c_t, is passed through a tanh function and multiplied by the output gate output, o_t, to calculate the final hidden state, h_t. This means that the learned parameters on the output gate effectively control which elements of the previous hidden state and current output are combined with the final cell state to carry over to the next time step as the new hidden state.

In our forward pass, we simply iterate through the model, initializing our hidden state and cell state and updating them at each time step using the LSTM cells until we are left with a final hidden state, which is output to the next layer of our neural network. By backpropagating through all the layers of our LSTM, we can calculate the gradients relative to the loss of the network and so we know which direction to update our parameters through gradient descent. We get several matrices or parameters—one for the input gate, one for the output gate, and one for the forget gate.

Because we get more parameters than for a simple RNN and our computation graph is more complex, the process of backpropagating through the network and updating the weights will likely take longer than for a simple RNN. However, despite the longer training time, we have shown that LSTM offers significant advantages over a conventional RNN as the output gate, input gate, and forget gate all combine to give the model the ability to determine which elements of the input should be used to update the hidden state and which elements of the hidden state should be forgotten going forward, which means the model is better able to form long-term dependencies and retain information from previous sequence steps.

Bidirectional LSTMs

We previously mentioned that a downside of simple RNNs is that they fail to capture the full context of a word within a sentence as they are backward-looking only. At each time step of the RNN, only the previously seen words are considered and the words occurring next within the sentence are not taken into account. While basic LSTMs are similarly backward-facing, we can use a modified version of LSTM, known as a **bidirectional LSTM**, which considers both the words before and after it at each time step within the sequence.

Bidirectional LSTMs process sequences in regular order and reverse order simultaneously, maintaining two hidden states. We'll call the forward hidden state f_t and use r_t for the reverse hidden state:

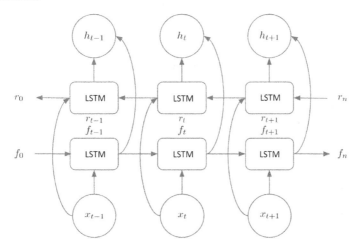

Figure 5.10 – The bidirectional LSTM process

Here, we can see that we maintain these two hidden states throughout the whole process and use them to calculate a final hidden state, h_t. Therefore, if we wish to calculate the final hidden state at time step t, we use the forward hidden state, f_t, which has seen all words up to and including input x_t, as well as the reverse hidden state, r_t, which has seen all the words after and including x_t. Therefore, our final hidden state, h_t, comprises hidden states that have seen all the words in the sentence, not just the words occurring before time step t. This means that the context of any given word within the whole sentence can be better captured. Bidirectional LSTMs have proven to offer improved performance on several NLP tasks over conventional unidirectional LSTMs.

Building a sentiment analyzer using LSTMs

We will now look at how to build our own simple LSTM to categorize sentences based on their sentiment. We will train our model on a dataset of 3,000 reviews that have been categorized as positive or negative. These reviews come from three different sources—film reviews, product reviews, and location reviews—in order to ensure that our sentiment analyzer is robust. The dataset is balanced so that it consists of 1,500 positive reviews and 1,500 negative reviews. We will start by importing our dataset and examining it:

```
with open("sentiment labelled sentences/sentiment.txt") as f:
    reviews = f.read()

data = pd.DataFrame([review.split('\t') for review in
                    reviews.split('\n')])

data.columns = ['Review','Sentiment']

data = data.sample(frac=1)
```

This returns the following output:

	Review	Sentiment
28	It deserves strong love.	1
1407	Food was really good and I got full petty fast.	1
2967	Beautiful styling though.	1
2807	The reception through this headset is excellent.	1
2616	Poor product.	0

Figure 5.11 – Output of the dataset

We read in our dataset from the file. Our dataset is tab-separated, so we split it up with tabs and the new line character. We rename our columns and then use the sample function to randomly shuffle our data. Looking at our dataset, the first thing we need to be able to do is preprocess our sentences to feed them into our LSTM model.

Preprocessing the data

First, we create a function to tokenize our data, splitting each review into a list of individual preprocessed words. We loop through our dataset and for each review, we remove any punctuation, convert letters into lowercase, and remove any trailing whitespace. We then use the NLTK tokenizer to create individual tokens from this preprocessed text:

```
def split_words_reviews(data):
    text = list(data['Review'].values)
    clean_text = []
    for t in text:
        clean_text.append(t.translate(str.maketrans('', '',
                    punctuation)).lower().rstrip())
    tokenized = [word_tokenize(x) for x in clean_text]
    all_text = []
    for tokens in tokenized:
        for t in tokens:
            all_text.append(t)
    return tokenized, set(all_text)
reviews, vocab = split_words_reviews(data)
reviews[0]
```

This results in the following output:

```
['great', 'time', 'family', 'dinner', 'on', 'a', 'sunday', 'night']
```

Figure 5.12 – Output of NTLK tokenization

We return the reviews themselves, as well as a set of all words within all the reviews (that is, the vocabulary/corpus), which we will use to create our vocab dictionaries.

In order to fully prepare our sentences for entry into a neural network, we must convert our words into numbers. In order to do this, we create a couple of dictionaries, which will allow us to convert data from word into index and from index into word. To do this, we simply loop through our corpus and assign an index to each unique word:

```
def create_dictionaries(words):
    word_to_int_dict = {w:i+1 for i, w in enumerate(words)}
    int_to_word_dict = {i:w for w, i in word_to_int_dict.
                        items()}
    return word_to_int_dict, int_to_word_dict


word_to_int_dict, int_to_word_dict = create_dictionaries(vocab)


int_to_word_dict
```

This gives the following output:

```
{1: 'tonight',
 2: 'attractive',
 3: 'magic',
 4: 'acknowledged',
 5: 'inexpensive',
 6: 'receiving',
 7: 'expert',
 8: 'nervous',
 9: 'muffled',
 10: 'appalling',
```

Figure 5.13 – Assigning an index to each word

Our neural network will take input of a fixed length; however, if we explore our reviews, we will see that our reviews are all of different lengths. In order to ensure that all of our inputs are of the same length, we will *pad* our input sentences. This essentially means that we add empty tokens to shorter sentences so that all the sentences are of the same length. We must first decide on the length of the padding we wish to implement. We first calculate the maximum length of a sentence in our input reviews, as well as the average length:

```
print(np.max([len(x) for x in reviews]))
print(np.mean([len(x) for x in reviews]))
```

This gives the following:

```
70
11.783666666666667
```

Figure 5.14 – Length value

We can see that the longest sentence is 70 words long and the average sentence length has a length of 11.78. To capture all the information from all our sentences, we want to pad all of our sentences so that they have a length of 70. However, using longer sentences means longer sequences, which causes our LSTM layer to become deeper. This means model training takes longer as we have to backpropagate our gradients through more layers, but it also means that a large percentage of our inputs would just be sparse and full of empty tokens, which makes learning from our data much less efficient. This is illustrated by the fact that our maximum sentence length is much larger than our average sentence length. In order to capture the majority of our sentence information without unnecessarily padding our inputs and making them too sparse, we opt to use an input size of 50. You may wish to experiment with using different input sizes between 20 and 70 to see how this affects your model performance.

We will create a function that allows us to pad our sentences so that they are all the same size. For reviews shorter than the sequence length, we pad them with empty tokens. For reviews longer than the sequence length, we simply drop any tokens over the maximum sequence length:

```python
def pad_text(tokenized_reviews, seq_length):

    reviews = []

    for review in tokenized_reviews:
        if len(review) >= seq_length:
            reviews.append(review[:seq_length])
        else:
            reviews.append([''] * (seq_length-len(review)) +
                review)

    return np.array(reviews)

padded_sentences = pad_text(reviews, seq_length = 50)

padded_sentences[0]
```

Our padded sentence looks like this:

```
array(['', '', '', '', '', '', '', '', '', '', '', '', '', '', '', '',
       '', '', '', '', '', '', '', '', '', '', '', '', '', '', '', '',
       '', '', '', '', '', '', '', '', 'great', 'time', 'family',
       'dinner', 'on', 'a', 'sunday', 'night'], dtype='<U33')
```

Figure 5.15 – Padding the sentences

We must make one further adjustment to allow the use of empty tokens within our model. Currently, our vocabulary dictionaries do not know how to convert empty tokens into integers to use within our network. Because of this, we manually add these to our dictionaries with index 0, which means empty tokens will be given a value of 0 when fed into our model:

```
int_to_word_dict[0] = ''
word_to_int_dict[''] = 0
```

We are now very nearly ready to begin training our model. We perform one final step of preprocessing and encode all of our padded sentences as numeric sequences for feeding into our neural network. This means that the previous padded sentence now looks like this:

```
encoded_sentences = np.array([[word_to_int_dict[word] for word
in review] for review in padded_sentences])
encoded_sentences[0]
```

Our encoded sentence is represented as follows:

```
array([   0,    0,    0,    0,    0,    0,    0,    0,    0,    0,    0,
          0,    0,    0,    0,    0,    0,    0,    0,    0,    0,    0,
          0,    0,    0,    0,    0,    0,    0,    0,    0,    0,    0,
          0,    0,    0,    0,    0,    0,    0,    0, 3869,  643,
       4472, 3286, 3868, 3218, 1261,  505])
```

Figure 5.16 – Encoding the sentence

Now that we have all our input sequences encoded as numerical vectors, we are ready to begin designing our model architecture.

Model architecture

Our model will consist of several main parts. Besides the input and output layers that are common to many neural networks, we will first require an **embedding layer**. This is so that our model learns the vector representations of the words it is being trained on. We could opt to use precomputed embeddings (such as GLoVe), but for demonstrative purposes, we will train our own embedding layer. Our input sequences are fed through the input layer and come out as sequences of vectors.

These vector sequences are then fed into our **LSTM layer**. As explained in detail earlier in this chapter, the LSTM layer learns sequentially from our sequence of embeddings and outputs a single vector output representing the final hidden state of the LSTM layer. This final hidden state is finally passed through a further **hidden layer** before the final output node predicts a value between 0 and 1, indicating whether the input sequence was a positive or negative review. This means that our model architecture looks something like this:

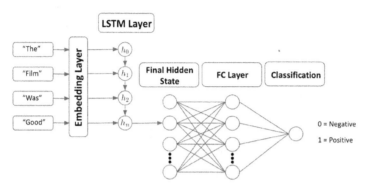

Figure 5.17 – Model architecture

We will now demonstrate how to code this model from scratch using PyTorch. We create a class called SentimentLSTM, which inherits from the nn.Module class. We define our init parameters as the size of our vocab, the number of LSTM layers our model will have, and the size of our model's hidden state:

```
class SentimentLSTM(nn.Module):

    def __init__(self, n_vocab, n_embed, n_hidden, n_output,
    n_layers, drop_p = 0.8):
        super().__init__()

        self.n_vocab = n_vocab
        self.n_layers = n_layers
        self.n_hidden = n_hidden
```

We then define each of the layers of our network. Firstly, we define our embedding layer, which will have the length of the number of words in our vocabulary and the size of the embedding vectors as a n_embed hyperparameter to be specified. Our LSTM layer is defined using the output vector size from the embedding layer, the length of the model's hidden state, and the number of layers that our LSTM layer will have. We also add an argument to specify that our LSTM can be trained on batches of data and an argument to allow us to implement network regularization via dropout. We define a further dropout layer with probability, drop_p (a hyperparameter to be specified on model creation), as well as our definitions of our final fully connected layer and output/prediction node (with a sigmoid activation function):

```python
self.embedding = nn.Embedding(n_vocab, n_embed)
self.lstm = nn.LSTM(n_embed, n_hidden, n_layers,
                    batch_first = True, dropout = drop_p)
self.dropout = nn.Dropout(drop_p)
self.fc = nn.Linear(n_hidden, n_output)
self.sigmoid = nn.Sigmoid()
```

Next, we need to define our forward pass within our model class. Within this forward pass, we just chain together the output of one layer to become the input into our next layer. Here, we can see that our embedding layer takes input_words as input and outputs the embedded words. Then, our LSTM layer takes embedded words as input and outputs lstm_out. The only nuance here is that we use view() to reshape our tensors from the LSTM output to be the correct size for input into our fully connected layer. The same also applies for reshaping the output of our hidden layer to match that of our output node. Note that our output will return a prediction for class = 0 and class = 1, so we slice the output to only return a prediction for class = 1—that is, the probability that our sentence is positive:

```python
def forward (self, input_words):

    embedded_words = self.embedding(input_words)
    lstm_out, h = self.lstm(embedded_words)
    lstm_out = self.dropout(lstm_out)
    lstm_out = lstm_out.contiguous().view(-1,
                            self.n_hidden)
    fc_out = self.fc(lstm_out)
    sigmoid_out = self.sigmoid(fc_out)
    sigmoid_out = sigmoid_out.view(batch_size, -1)
```

```
        sigmoid_last = sigmoid_out[:, -1]

        return sigmoid_last, h
```

We also define a function called `init_hidden()`, which initializes our hidden layer with the dimensions of our batch size. This allows our model to train and predict on many sentences at once, rather than just training on one sentence at a time, if we so wish. Note that we define `device` as `"cpu"` here to run it on our local processor. However, it is also possible to set this to a CUDA-enabled GPU in order to train it on a GPU if you have one:

```
    def init_hidden (self, batch_size):

        device = "cpu"
        weights = next(self.parameters()).data
        h = (weights.new(self.n_layers, batch_size,\
                self.n_hidden).zero_().to(device),\
            weights.new(self.n_layers, batch_size,\
                self.n_hidden).zero_().to(device))

        return h
```

We then initialize our model by creating a new instance of the `SentimentLSTM` class. We pass the size of our vocab, the size of our embeddings, the size of our hidden state, as well as the output size, and the number of layers in our LSTM:

```
n_vocab = len(word_to_int_dict)
n_embed = 50
n_hidden = 100
n_output = 1
n_layers = 2

net = SentimentLSTM(n_vocab, n_embed, n_hidden, n_output,
n_layers)
```

Now that we have defined our model architecture fully, it's time to begin training our model.

Training the model

To train our model, we must first define our datasets. We will train our model using a training set of data, evaluate our trained model at each step on a validation set, and then finally, measure our model's final performance using an unseen test set of data. The reason we use a test set that is separate from our validation training is that we may wish to fine-tune our model hyperparameters based on the loss against the validation set. If we do this, we may end up picking the hyperparameters that are only optimal in performance for that particular validation set of data. We evaluate a final time against an unseen test set to make sure our model generalizes well to data it hasn't seen before at any part of the training loop.

We have already defined our model inputs (*x*) as `encoded_sentences`, but we must also define our model output (*y*). We do this simply, as follows:

```
labels = np.array([int(x) for x in data['Sentiment'].values])
```

Next, we define our training and validation ratios. In this case, we will train our model on 80% of the data, validate on a further 10% of the data, and finally, test on the remaining 10% of the data:

```
train_ratio = 0.8
valid_ratio = (1 - train_ratio)/2
```

We then use these ratios to slice our data and transform them into tensors and then tensor datasets:

```
total = len(encoded_sentences)
train_cutoff = int(total * train_ratio)
valid_cutoff = int(total * (1 - valid_ratio))


train_x, train_y = torch.Tensor(encoded_sentences[:train_
cutoff]).long(), torch.Tensor(labels[:train_cutoff]).long()

valid_x, valid_y = torch.Tensor(encoded_sentences[train_cutoff
: valid_cutoff]).long(), torch.Tensor(labels[train_cutoff :
valid_cutoff]).long()
test_x, test_y = torch.Tensor(encoded_sentences[valid_
cutoff:]).long(), torch.Tensor(labels[valid_cutoff:])


train_data = TensorDataset(train_x, train_y)
valid_data = TensorDataset(valid_x, valid_y)
test_data = TensorDataset(test_x, test_y)
```

Then, we use these datasets to create PyTorch `DataLoader` objects. `DataLoader` allows us to batch process our datasets with the `batch_size` parameter, allowing different batch sizes to be easily passed to our model. In this instance, we will keep it simple and set `batch_size = 1`, which means our model will be trained on individual sentences, rather than using larger batches of data. We also opt to randomly shuffle our `DataLoader` objects so that data is passed through our neural network in random order, rather than the same order each epoch, potentially removing any biased results from the training order:

```
batch_size = 1

train_loader = DataLoader(train_data, batch_size = batch_size,
                          shuffle = True)
valid_loader = DataLoader(valid_data, batch_size = batch_size,
                          shuffle = True)
test_loader = DataLoader(test_data, batch_size = batch_size,
                         shuffle = True)
```

Now that we have defined our `DataLoader` object for each of our three datasets, we define our training loop. We first define a number of hyperparameters, which will be used within our training loop. Most importantly, we define our loss function as binary cross entropy (as we are dealing with predicting a single binary class) and we define our optimizer to be `Adam` with a learning rate of `0.001`. We also define our model to run for a short number of epochs (to save time) and set `clip = 5` to define our gradient clipping:

```
print_every = 2400
step = 0
n_epochs = 3
clip = 5
criterion = nn.BCELoss()
optimizer = optim.Adam(net.parameters(), lr = 0.001)
```

The main body of our training loop looks like this:

```
for epoch in range(n_epochs):
    h = net.init_hidden(batch_size)


    for inputs, labels in train_loader:
        step += 1
```

```
net.zero_grad()
output, h = net(inputs)
loss = criterion(output.squeeze(), labels.float())
loss.backward()
nn.utils.clip_grad_norm(net.parameters(), clip)
optimizer.step()
```

Here, we just train our model for a number of epochs, and for every epoch, we first initialize our hidden layer using the batch size parameter. In this instance, we set `batch_size = 1` as we are just training our model one sentence at a time. For each batch of input sentences and labels within our train loader, we first zero our gradients (to stop them accumulating) and calculate our model outputs using the forward pass of our data using the model's current state. Using this output, we then calculate our loss using the predicted output from the model and the correct labels. We then perform a backward pass of this loss through our network to calculate the gradients at each stage. Next, we use the `grad_clip_norm()` function to clip our gradients as this will stop our gradients from exploding, as mentioned earlier in this chapter. We defined `clip = 5`, meaning the maximum gradient at any given node is 5. Finally, we update our weights using the gradients calculated on our backward pass by calling `optimizer.step()`.

If we run this loop by itself, we will train our model. However, we want to evaluate our model performance after every epoch in order to determine its performance on a validation set of data. We do this as follows:

```
if (step % print_every) == 0:
            net.eval()
            valid_losses = []

            for v_inputs, v_labels in valid_loader:

                v_output, v_h = net(v_inputs)
                v_loss = criterion(v_output.squeeze(),
                                    v_labels.float())
                valid_losses.append(v_loss.item())

            print("Epoch: {}/{}".format((epoch+1), n_epochs),
                "Step: {}".format(step),
                "Training Loss: {:.4f}".format(loss.item()),
```

```
              "Validation Loss: {:.4f}".format(np.
                           mean(valid_losses)))
    net.train()
```

This means that at the end of each epoch, our model calls `net.eval()` to freeze the weights of our model and performs a forward pass using our data as before. Note that dropout is also not applied when we are in evaluation mode. However, this time, instead of using the training data loader, we use the validation loader. By doing this, we can calculate the total loss of the model's current state over our validation set of data. Finally, we print our results and call `net.train()` to unfreeze our model's weights so that we can train again on the next epoch. Our output looks something like this:

```
Epoch: 1/3 Step: 2400 Training Loss: 0.2800 Validation Loss: 0.7575
Epoch: 2/3 Step: 4800 Training Loss: 0.0129 Validation Loss: 0.6676
Epoch: 3/3 Step: 7200 Training Loss: 0.0673 Validation Loss: 0.6420
```

Figure 5.18 – Training the model

Finally, we can save our model for future use:

```
torch.save(net.state_dict(), 'model.pkl')
```

After training our model for three epochs, we notice two main things. We'll start with the good news first—our model is learning something! Not only has our training loss fallen, but we can also see that our loss on the validation set has fallen after each epoch. This means that our model is better at predicting sentiment on an unseen set of data after just three epochs! The bad news, however, is that our model is massively overfitting. Our training loss is much lower than that of our validation loss, showing that while our model has learned how to predict the training set of data very well, this doesn't generalize as well to an unseen set of data. This was expected to happen as we are using a very small set of training data (just 2,400 training sentences). As we are training a whole embedding layer, it is possible that many of the words occur just once in the training set and never in the validation set and vice versa, making it practically impossible for the model to generalize all the different variety of words within our corpus. In practice, we would hope to train our model on a much larger dataset to allow our model to learn how to generalize much better. We have also trained this model over a very short time period and have not performed hyperparameter tuning to determine the best possible iteration of our model. Feel free to try changing some of the parameters within the model (such as the training time, hidden state size, embedding size, and so on) in order to improve the performance of the model.

Although our model overfitted, it has still learned something. We now wish to evaluate our model on a final test set of data. We perform one final pass on the data using the test loader we defined earlier. Within this pass, we loop through all of our test data and make predictions using our final model:

```
net.eval()
test_losses = []
num_correct = 0


for inputs, labels in test_loader:

    test_output, test_h = net(inputs)
    loss = criterion(test_output, labels)
    test_losses.append(loss.item())

    preds = torch.round(test_output.squeeze())
    correct_tensor = preds.eq(labels.float().view_as(preds))
    correct = np.squeeze(correct_tensor.numpy())
    num_correct += np.sum(correct)

print("Test Loss: {:.4f}".format(np.mean(test_losses)))
print("Test Accuracy: {:.2f}".format(num_correct/len(test_
loader.dataset)))
```

Our performance on our test set of data is as follows:

Test Loss: 0.6598
Test Accuracy: 0.76

Figure 5.19 – Output values

We then compare our model predictions with our true labels to get `correct_tensor`, which is a vector that evaluates whether each of our model's predictions was correct. We then sum this vector and divide it by its length to get our model's total accuracy. Here, we get an accuracy of 76%. While our model is certainly far from perfect, given our very small training set and limited training time, this is not bad at all! This just serves to illustrate how useful LSTMs can be when it comes to learning from NLP data. Next, we will show how we can use our model to make predictions from new data.

Using our model to make predictions

Now that we have a trained model, it should be possible to repeat our preprocessing steps on a new sentence, pass this into our model, and get a prediction of it's sentiment. We first create a function to preprocess our input sentence to predict:

```python
def preprocess_review(review):
    review = review.translate(str.maketrans('', '',
                    punctuation)).lower().rstrip()
    tokenized = word_tokenize(review)
    if len(tokenized) >= 50:
        review = tokenized[:50]
    else:
        review= ['0']*(50-len(tokenized)) + tokenized

    final = []

    for token in review:
        try:
            final.append(word_to_int_dict[token])

        except:
            final.append(word_to_int_dict[''])

    return final
```

We remove punctuation and trailing whitespace, convert letters into lowercase, and tokenize our input sentence as before. We pad our sentence to a sequence with a length of 50 and then convert our tokens into numeric values using our precomputed dictionary. Note that our input may contain new words that our network hasn't seen before. In this case, our function treats these as empty tokens.

Next, we create our actual `predict()` function. We preprocess our input review, convert it into a tensor, and pass this into a data loader. We then loop through this data loader (even though it only contains one sentence) and pass our review through our network to obtain a prediction. Finally, we evaluate our prediction and print whether it is a positive or negative review:

```python
def predict(review):
    net.eval()
```

```
words = np.array([preprocess_review(review)])
padded_words = torch.from_numpy(words)
pred_loader = DataLoader(padded_words, batch_size = 1,
                         shuffle = True)

for x in pred_loader:
    output = net(x)[0].item()


msg = "This is a positive review." if output >= 0.5 else
      "This is a negative review."
print(msg)
print('Prediction = ' + str(output))
```

Finally, we just call `predict()` on our review to make a prediction:

```
predict("The film was good")
```

This results in the following output:

```
This is a positive review.
Prediction = 0.917565107345581
```

Figure 5.20 – Prediction string on a positive value

We also try using `predict()` on the negative value:

```
predict("It was not good")
```

This results in the following output:

```
This is a negative review.
Prediction = 0.2955784499645233
```

Figure 5.21 – Prediction string on a negative value

We have now built an LSTM model to perform sentiment analysis from the ground up. Although our model is far from perfect, we have demonstrated how we can take some sentiment labeled reviews and train a model to be able to make predictions on new reviews. Next, we will show how we can host our model on the Heroku cloud platform so that other people can make predictions using your model

Deploying the application on Heroku

We have now trained our model on our local machine and we can use this to make predictions. However, this isn't necessarily any good if you want other people to be able to use your model to make predictions. If we host our model on a cloud-based platform, such as Heroku, and create a basic API, other people will be able to make calls to the API to make predictions using our model.

Introducing Heroku

Heroku is a cloud-based platform where you can host your own basic programs. While the free tier of Heroku has a maximum upload size of 500 MB and limited processing power, this should be sufficient for us to host our model and create a basic API in order to make predictions using our model.

The first step is to create a free account on Heroku and install the Heroku app. Then, in the command line, type the following command:

```
heroku login
```

Log in using your account details. Then, create a new heroku project by typing the following command:

```
heroku create sentiment-analysis-flask-api
```

Note that all the project names must be unique, so you will need to pick a project name that isn't sentiment-analysis-flask-api.

Our first step is building a basic API using Flask.

Creating an API using Flask – file structure

Creating an API is fairly simple using Flask as Flask contains a default template required to make an API:

First, in the command line, create a new folder for your flask API and navigate to it:

```
mkdir flaskAPI
cd flaskAPI
```

Then, create a virtual environment within the folder. This will be the Python environment that your API will use:

```
python3 -m venv vir_env
```

Within your environment, install all the packages that you will need using `pip`. This includes all the packages that you use within your model program, such as NLTK, `pandas`, NumPy, and PyTorch, as well as the packages you will need to run the API, such as Flask and Gunicorn:

```
pip install nltk pandas numpy torch flask gunicorn
```

We then create a list of requirements that our API will use. Note that when we upload this to Heroku, Heroku will automatically download and install all the packages within this list. We can do this by typing the following:

```
pip freeze > requirements.txt
```

One adjustment we need to make is to replace the `torch` line within the `requirements.txt` file with the following:

```
https://download.pytorch.org/whl/cpu/torch-1.3.1%2Bcpu-cp37-
cp37m-linux_x86_64.whl
```

This is a link to the wheel file of the version of PyTorch that only contains the CPU implementation. The full version of PyTorch that includes full GPU support is over 500 MB in size, so it will not run on the free Heroku cluster. Using this more compact version of PyTorch means that you will still be able to run your model using PyTorch on Heroku. Finally, we create three more files within our folder, as well as a final directory for our models:

```
touch app.py
touch Procfile
touch wsgi.py
mkdir models
```

Now, we have created all the files we will need for our Flash API and we are ready to start making adjustments to our file.

Creating an API using Flask – API file

Within our app.py file, we can begin building our API:

1. We first carry out all of our imports and create a predict route. This allows us to call our API with the predict argument in order to run a predict() method within our API:

```
import flask
from flask import Flask, jsonify, request
import json

import pandas as pd
from string import punctuation
import numpy as np
import torch
from nltk.tokenize import word_tokenize
from torch.utils.data import TensorDataset, DataLoader
from torch import nn
from torch import optim

app = Flask(__name__)
@app.route('/predict', methods=['GET'])
```

2. Next, we define our predict() method within our app.py file. This is largely a rehash of our model file, so to avoid repetition of code, it is advised that you look at the completed app.py file within the GitHub repository linked in the *Technical requirements* section of this chapter. You will see that there are a few additional lines. Firstly, within our preprocess_review() function, we will see the following lines:

```
with open('models/word_to_int_dict.json') as handle:
    word_to_int_dict = json.load(handle)
```

This takes the word_to_int dictionary we computed within our main model notebook and loads it into our model. This is so that our word indexing is consistent with our trained model. We then use this dictionary to convert our input text into an encoded sequence. Be sure to take the word_to_int_dict.json file from the original notebook output and place it within the models directory.

3. Similarly, we must also load the weights from our trained model. We first define our `SentimentLSTM` class and the load our weights using `torch.load`. We will use the `.pkl` file from our original notebook, so be sure to place this in the `models` directory as well:

```
model = SentimentLSTM(5401, 50, 100, 1, 2)
model.load_state_dict(torch.load("models/model_nlp.pkl"))
```

4. We must also define the input and outputs of our API. We want our model to take the input from our API and pass this to our `preprocess_review()` function. We do this using `request.get_json()`:

```
request_json = request.get_json()
i = request_json['input']
words = np.array([preprocess_review(review=i)])
```

5. To define our output, we return a JSON response consisting of the output from our model and a response code, `200`, which is what is returned by our predict function:

```
output = model(x)[0].item()
response = json.dumps({'response': output})
        return response, 200
```

6. With the main body of our app complete, there are just two more additional things we must add in order to make our API run. We must first add the following to our `wsgi.py` file:

```
from app import app as application
if __name__ == "__main__":
    application.run()
```

7. Finally, add the following to our Procfile:

```
web: gunicorn app:app --preload
```

That's all that's required for the app to run. We can test that our API runs by first starting the API locally using the following command:

```
gunicorn --bind 0.0.0.0:8080 wsgi:application -w 1
```

Once the API is running locally, we can make a request to the API by passing it a sentence to predict the outcome:

```
curl -X GET http://0.0.0.0:8080/predict -H "Content-Type:
application/json" -d '{"input":"the film was good"}'
```

If everything is working correctly, you should receive a prediction from the API. Now that we have our API making predictions locally, it is time to host it on Heroku so that we can make predictions in the cloud.

Creating an API using Flask – hosting on Heroku

We first need to commit our files to Heroku in a similar way to how we would commit files using GitHub. We define our working flaskAPI directory as a git folder by simply running the following command:

```
git init
```

Within this folder, we add the following code to the .gitignore file, which will stop us from adding unnecessary files to the Heroku repo:

```
vir_env
__pycache__/
.DS_Store
```

Finally, we add our first commit function and push it to our heroku project:

```
git add . -A
git commit -m 'commit message here'
git push heroku master
```

This may take some time to compile as not only does the system have to copy all the files from your local directory to Heroku, but Heroku will automatically build your defined environment, installing all the required packages and running your API.

Now, if everything has worked correctly, your API will automatically run on the Heroku cloud. In order to make predictions, you can simply make a request to the API using your project name instead of sentiment-analysis-flask-api:

```
curl -X GET https://sentiment-analysis-flask-api.herokuapp.com/
predict -H "Content-Type: application/json" -d '{"input":"the
film was good"}'
```

Your application will now return a prediction from the model. Congratulations, you have now learned how to train an LSTM model from scratch, upload it to the cloud, and make predictions using it! Going forward, this tutorial will hopefully serve as a basis for you to train your own LSTM models and deploy them to the cloud yourself.

Summary

In this chapter, we discussed the fundamentals of RNNs and one of their main variations, LSTM. We then demonstrated how you can build your own RNN from scratch and deploy it on the cloud-based platform Heroku. While RNNs are often used for deep learning on NLP tasks, they are by no means the only neural network architecture suitable for this task.

In the next chapter, we will look at convolutional neural networks and show how they can be used for NLP learning tasks.

6
Convolutional Neural Networks for Text Classification

In the previous chapter, we showed how RNNs can be used to provide sentiment classifications for text. However, RNNs are not the only neural network architecture that can be used for NLP classification tasks. **Convolutional neural networks (CNNs)** are another such architecture.

RNNs rely on sequential modeling, maintain a hidden state, and then step sequentially through the text word by word, updating the state at each iteration. CNNs do not rely on the sequential element of language, but instead try and learn from the text by perceiving each word in the sentence individually and learning its relationship to the words surrounding it within the sentence.

While CNNs are more commonly used for classifying images for the reasons mentioned here, they have been shown to be effective at classifying text as well. While we do perceive text as a sequence, we also know that the meaning of individual words in the sentence depends on their context and the words they appear next to. Although CNNs and RNNs learn from text in different ways, they have both shown to be effective in text classification, and which one to use in any given situation depends on the nature of the task.

In this chapter, we will explore the basic theory behind CNNs, as well as construct a CNN from scratch that will be used to classify text. We will cover the following topics:

- Exploring CNNs
- Building a CNN for text classification

Let's get started!

Technical requirements

All the code for this chapter can be found at `https://github.com/PacktPublishing/Hands-On-Natural-Language-Processing-with-PyTorch-1.x`.

Exploring CNNs

The basis for CNNs comes from the field of computer vision but can conceptually be extended to work on NLP as well. The way the human brain processes and understands images is not on a pixel-by-pixel basis, but as a holistic map of an image and how each part of the image relates to the other parts.

A good analogy of CNNs would be how the human mind processes a picture versus how it processes a sentence. Consider the sentence, *This is a sentence about a cat.* When you read that sentence you read the first word, followed by the second word and so forth. Now, consider a picture of a cat. It would be foolish to assimilate the information within the picture by looking at the first pixel, followed by the second pixel. Instead, when we look at something, we perceive the whole image at once, rather than as a sequence.

For example, if we take a black and white representation of an image (in this case, the digit 1), we can see that we can transform this into a vector representation, where the color of each pixel is denoted by a 0 or a 1:

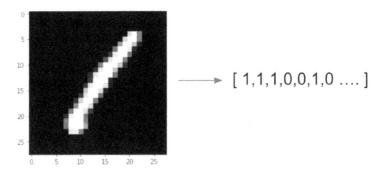

Figure 6.1 – Vector representation of an image

However, if we think about this in machine learning terms and treat this vector as features of our model, does the fact that any single pixel is black or white make it more or less likely that the picture is of a given digit? Does a white pixel in the top-right corner make the picture more likely to be a four or a seven? Imagine if we were trying to detect something more complex, such as whether a picture is of a dog or a cat. Does a brown pixel in the middle of the screen make the photo more likely to be a cat or a dog? Intuitively, we see that individual pixel values do not mean a lot when it comes to image classification. However, what we are interested in is the pixel's relationships to one another.

In our case of digit representation, we know that a long vertical line is very likely to be a one and that any photo with a closed loop in it is more likely to be a zero, six, eight, or nine. By identifying and learning from patterns within our images, rather than just looking at individual pixels, we can better understand and identify these images. This is exactly what CNNs aim to achieve.

Convolutions

The basic concept behind CNNs is that of convolutions. A **convolution** is essentially a sliding window function that's applied to a matrix in order to capture information from the surrounding pixels. In the following diagram, we can see an example of convolution in action:

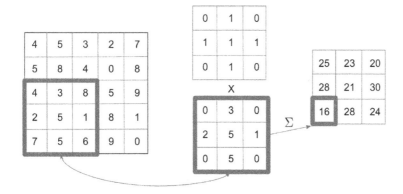

Figure 6.2 – Convolution in action

On the left, we have the image that we are processing, while at the top, we have the convolution kernel we wish to apply. For every 3x3 block within our image, we multiply this by our kernel to give us our convolution matrix at the bottom. We then sum (or average) the convolved matrix to get our single output value for this 3x3 block within our initial image. Note that within our 5x5 initial images, there are nine possible 3x3 blocks we can overlay onto this. As we apply this convolution process for every 3x3 block within our initial image, we are left with a final processed convolution that is 3x3.

In a large image (or a complex sentence, in the case of NLP), we will also need to implement pooling layers. In our preceding example, applying a 3x3 convolution to a 5x5 image results in a 3x3 output. However, if we applied a 3x3 convolution to a 100x100 pixel image, this would only reduce the output to 98x98. This hasn't reduced the dimensionality of the image enough to perform deep learning effectively (as we would have to learn 98x98 parameters per convolutional layer). Therefore, we apply a pooling layer to further reduce the dimensionality of the layer.

A pooling layer applies a function (typically, a max function) to the output of the convolutional layer in order to reduce its dimensionality. This function is applied over a sliding window, similar to how our convolutions are performed, except now our convolutions do not overlap. Let's assume our convolutional layer has an output of 4x4 and we apply a 2x2 max pooling function to our output. This means that for every smaller 2x2 grid within our layer, we apply a max function and keep the resulting output. We can see this in the following diagram:

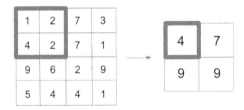

Figure 6.3 – Pooling layers

These pooling layers have been shown to effectively reduce the dimensionality of our data, while still retaining much of the essential information from the convolutional layer.

This combination of convolutions and pooling layers is essentially how CNNs learn from images. We can see that by applying many of these convolutional processes (also known as **convolutional layers**), we are able to capture information about any given pixel's relationship to its neighboring pixels. Within CNNs, the parameters we aim to learn are the values of the convolution kernel itself. This means that our model effectively learns how it should convolve across an image in order to be able to extract the necessary information required to make a classification.

There are two main advantages to using convolutions in this context. Firstly, we are able to compose a series of low-level features into a higher-level feature; that is, a 3x3 patch on our initial image is composed into a single value. This effectively acts as a form of feature reduction and allows us to only extract the relevant information from our image. The other advantage that using convolutions has is that it makes our model location invariant. In our digit detector example, we do not care if the number occurs on the right-hand side of the image or the left-hand side; we just want to be able to detect it. As our convolutions will detect specific patterns within our image (that is, edges), this makes our model location invariant as the same features will theoretically be picked up by the convolutions, no matter where they occur within the image.

While these principles are useful for understanding how convolutions work in image data, they can also be applied to NLP data. We'll look at this in the next section.

Convolutions for NLP

As we have seen many times in this book, we can represent individual words numerically as vectors, and represent whole sentences and documents as a sequence of vectors. When we represent our sentence as a sequence of vectors, we can represent this as a matrix. If we have a matrix representation of a given sentence, we notice immediately that this is similar to the image we convolved over within our image convolutions. Therefore, we can apply convolutions to NLP in a similar fashion to images, provided we can represent our text as a matrix.

Let's first consider the basis for using this methodology. When we looked at n-grams previously, we saw that the context of a word in a sentence depends on the words preceding it and the words coming after it. Therefore, if we can convolve over a sentence in a way that allows us to capture the relation of one word to the words around it, we can theoretically detect patterns in language and use this to better classify our sentences.

It is also worth noting that our method of convolution is slightly different to our convolution on images. In our image matrix, we wish to capture the context of a single pixel relative to those surrounding it, whereas in a sentence, we wish to capture the context of a whole word vector, relative to the other vectors around it. Therefore, in NLP, we want to perform our convolutions across whole word vectors, rather than within word vectors. This is demonstrated in the following diagram.

We first represent our sentence as individual word vectors:

"The"	4	5	3	2	7
"book"	5	8	4	0	8
"was"	4	3	8	5	9
"very"	2	5	1	8	1
"good"	7	5	6	9	0

Figure 6.4 – Word vectors

We then apply a (2 x n) convolution over the matrix (where n is the length of our word vectors; in this instance, $n = 5$). We can convolve four different times using a (2 x n) filter, which reduces down into four outputs. You will notice that this is similar to a bi-gram model, where there are four possible bi-grams within a five-word sentence:

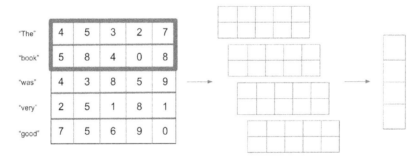

Figure 6.5 – Convolving word vectors into bi-grams

Similarly, we can do this for any number of n-grams; for example, $n=3$:

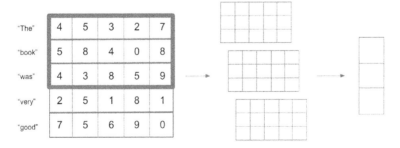

Figure 6.6 – Convolving word vectors into n-grams

One of the benefits of convolutional models such as this is there is no limit to the number of n-grams we can convolve over. We are also able to convolve over multiple different n-grams simultaneously. So, to capture both bi-grams and trigrams, we could set our model up like so:

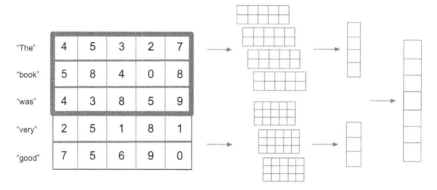

Figure 6.7 – Convolving word vectors into bi-grams and trigrams

Although CNNs for NLP have advantages such as those described in the preceding sections, they do have their drawbacks.

In CNNs for images, the assumption that a given pixel is likely related to those surrounding it is a reasonable one. When applied to NLP, while this assumption is partially correct, words can be semantically related, even if they are not in direct proximity to one another. A word at the start of a sentence can pertain to the word at the end of a sentence.

While our RNN models may be able to detect this relationship through longer-term memory dependencies, our CNNs may struggle as CNNs only capture the context of the words surrounding the target word.

That being said, CNNs for NLP have been proven to perform very well in some tasks, even though our language assumptions do not necessarily hold. Arguably, the main advantages of using CNNs for NLP are speed and efficiency. Convolutions can be easily implemented on GPUs, allowing for fast parallelized computation and training.

The way that relationships between words are captured is also much more efficient. In a true n-gram model, the model must learn individual representations for every single n-gram, whereas in our CNN models, we just learn the convolutional kernels, which will automatically extract the relationships between given word vectors.

Now that we have defined how our CNN will learn from our data, we can begin to code up a model from scratch.

Building a CNN for text classification

Now that we know the basics of CNNs, we can begin to build one from scratch. In the previous chapter, we built a model for sentiment prediction, where sentiment was a binary classifier; 1 for positive and 0 for negative. However, in this example, we will aim to build a CNN for **multi-class text classification**. In a multi-class problem, a particular example can only be classified as one of several classes. If an example can be classified as many different classes, then this is multi-label classification. Since our model is multi-class, this means that our model will aim at predicting which one of several classes our input sentence is classified as. While this problem is considerably more difficult than our binary classification task (as our sentence can now belong to one of many, rather than one of two classes), we will show that CNNs can deliver good performance on this task. We will first begin by defining our data.

Defining a multi-class classification dataset

In the previous chapter, we looked at a selection of reviews and learned a binary classification based on whether the review was positive or negative. For this task, we will look at data from the TREC (https://trec.nist.gov/data/qa.html) dataset, which is a common dataset used to evaluate the performance of a model's text-classification tasks. The dataset consists of a selection of questions, each of which falls into one of six broad semantic categories that our trained model will learn to classify. These six categories are as follows:

Label	Description	Example Question
ABBR	Abbreviations	What does BUD stand for?
DESC	Descriptions and abstract concepts	How did serfdom develop in and then leave Russia?
ENTY	Entities	What films featured the character Popeye Doyle?
HUM	Humans	What heavyweight boxer was known as The Wild Bull of the Pampas?
LOC	Locations	What U.S. state comes last in an alphabetical list?
NUM	Numerical Values	What is the highest number of home runs on record for any one game?

Figure 6.8 – Semantic categories in the TREC dataset

This means that unlike our previous classification class, where our model output was a single prediction between 0 and 1, our multi-class prediction model now returns a probability for each of the six possible classes. We assume that the prediction that's made is for the class with the highest prediction:

"ABBR"	"DESC"	"ENTY"	"HUM"	"LOC"	"NUM"
0.1	0.1	0.6	0.05	0.05	0.1

⟶ "ENTY"

Figure 6.9 – Prediction values

In this way, our model will now be able to perform classification tasks over several classes and we are no longer restricted to the 0 or 1 binary classification we looked at previously. Models with multiple classes may suffer in terms of predictions as there are more different classes to differentiate between.

In a binary classification model, assuming we had a balanced dataset, we would expect our model to have an accuracy of 50% if it were just to perform random guesses, whereas a multi-class model with five different classes would only have a baseline accuracy of 20%. This means that just because a multiclass model has an accuracy much lower than 100%, this does not mean the model itself is inherently bad at making predictions. This is particularly true when it comes to training models that predict from hundreds of different classes. In these cases, a model with just 50% accuracy would be considered to be performing very well.

Now that we have defined our multi-class classification problem, we need to load our data in order to train a model.

Creating iterators to load the data

In our LSTM model in the previous chapter, we simply used a `.csv` file containing all the data we used to train our model. We then manually converted this data into input tensors and fed them one by one into our network in order to train it. While this methodology is perfectly acceptable, it is not the most efficient one.

In our CNN model, we will instead look at creating data iterators from our data. These iterator objects allow us to easily generate small batches of data from our input data, thus allowing us to train our model using mini batches, rather than feeding our input data one by one into the network. This means that the gradients within our network are calculated across a whole batch of data and that parameter adjustments happen after each batch rather than after each individual row of data is passed through the network.

For our data, we will take our dataset from the TorchText package. This has the advantage of not only containing a number of datasets for model training, but it also allows us to easily tokenize and vectorize our sentences using the inbuilt functions.

Follow these steps:

1. We first import the data and dataset functions from TorchText:

   ```
   from torchtext import data
   from torchtext import datasets
   ```

2. Next, we create a field and label field we can use with the `TorchText` package. These define the initial processing steps our model will use to process our data:

   ```
   questions = data.Field(tokenize = 'spacy', batch_first =
   True)
   labels = data.LabelField(dtype = torch.float)
   ```

   Here, we set tokenize equal to `spacy` in order to set how our input sentences will be tokenized. `TorchText` then uses the `spacy` package to automatically tokenize the input sentences. `spacy` consists of an index of the English language, so any words are automatically transformed into the relevant tokens. You may need to install `spacy` in order for this to work. This can be done within the command line by typing the following:

   ```
   pip3 install spacy
   python3 -m spacy download en
   ```

   This installs `spacy` and downloads the English word index.

3. We also define the data type for our labels as floats, which will allow us to calculate our losses and gradients. After defining our fields, we can use these to split our input data. Using the `TREC` dataset from `TorchText`, we pass this our questions and labels fields in order to process the dataset accordingly. We then call the `split` function in order to automatically divide our dataset into a training set and a validation set:

   ```
   train_data, _ = datasets.TREC.splits(questions, labels)
   train_data, valid_data = train_data.split()
   ```

   Note that, normally, we can view our datasets in Python by simply calling the train data:

   ```
   train_data
   ```

However, here, we are dealing with a `TorchText` dataset object, rather than a dataset loaded into pandas, as we might be used to seeing. This means that our output from the preceding code is as follows:

```
<torchtext.data.dataset.Dataset at 0x1332b4e50>
```

Figure 6.10 – Output of the TorchText object

It is possible for us to view individual data within this dataset object; we just need to call the `.examples` parameter. Each of these examples will have a text and a label parameter that we can examine like so for the text:

```
train_data.examples[0].text
```

This returns the following output:

```
['What',
 'amount',
 'of',
 'money',
 'did',
 'the',
 'Philippine',
 'ex',
 '-',
 'dictator',
 'Marcos',
 'steal',
 'from',
 'the',
 'treasury',
 '?']
```

Figure 6.11 – Data in the dataset object

The label code is run as follows:

```
train_data.examples[0].label
```

This gives us the following output:

```
'NUM'
```

Figure 6.12 – Label of the dataset object

So, we can see that our input data consists of a tokenized sentence and that our label consists of the category that we wish to classify. We can also check the size of our training and validation sets, like so:

```
print(len(train_data))
print(len(valid_data))
```

This results in the following output:

3816

1636

Figure 6.13 – Sizes of the training and validation sets

This shows that our training to validation ratio is approximately 70% to 30%. It is worth noting exactly how our input sentence has been tokenized, namely that punctuation marks are treated as their own tokens.

Now that we know that our neural network will not take raw text as an input, we have to find some way of turning this into some form of embedding representation. While it is possible for us to train our own embedding layer, we can instead transform our data using the pre-computed `glove` vectors that we discussed in *Chapter 3, Performing Text Embeddings*. This also has the added benefit of making our model faster to train as we won't manually need to train our embedding layer from scratch:

```
questions.build_vocab(train_data,
                vectors = "glove.6B.200d",
                unk_init = torch.Tensor.normal_)


labels.build_vocab(train_data)
```

Here, we can see that by using the `build_vocab` function and passing our questions and labels as our training data, we can build a vocabulary composed of 200-dimensional GLoVe vectors. Note that the TorchText package will automatically download and grab the GLoVe vectors, so there is no need to manually install GLoVe in this instance. We also define how we wish to treat unknown values within our vocabulary (that is, what the model does if it is passed a token that isn't in the pretrained vocabulary). In this instance, we choose to treat them as a normal tensor with an unspecified value, although we will update this later.

We can now see that our vocabulary consists of a series of pre-trained 200-dimensional GLoVe vectors by calling the following command:

```
questions.vocab.vectors
```

This results in the following output:

```
tensor([[-0.5928,  1.9557, -0.4180,  ..., -0.1732, -1.0009, -0.7655],
        [-1.1752, -1.5180,  1.7845,  ...,  0.4673, -1.0432,  0.5888],
        [ 0.3911,  0.4019, -0.1505,  ..., -0.0348,  0.0798,  0.5031],
        ...,
        [-0.4949, -0.1262, -1.1698,  ...,  0.5565,  0.5634,  0.5782],
        [ 0.5741, -0.4343, -0.1119,  ...,  0.7629,  0.3831, -0.1570],
        [-0.2597,  0.6716,  0.5353,  ...,  1.6765,  0.3301, -0.1003]]])
```

Figure 6.14 – Tensor contents

Next, we create our data iterators. We create separate iterators for both our training and validation data. We first specify a device so that we are able to train our model faster using a CUDA-enabled GPU, if one is available. Within our iterators, we also specify the size of our batches to be returned by the iterators, which in this case is 64. You may wish to experiment with using different batch sizes for your model as this may affect training speed and how fast the model converges into its global optimum:

```
device = torch.device('cuda' if torch.cuda.is_available() else
                       'cpu')
train_iterator, valid_iterator = data.BucketIterator.splits(
    (train_data, valid_data),
    batch_size = 64,
    device = device)
```

Constructing the CNN model

Now that we have loaded the data, we are ready to create the model. We will use the following steps to do so:

1. We wish to build the structure of our CNN. We begin as usual by defining our model as a class that inherits from nn.Module:

```
class CNN(nn.Module):
    def __init__(self, vocab_size, embedding_dim,
    n_filters, filter_sizes, output_dim, dropout,
    pad_idx):

        super().__init__()
```

2. Our model is initialized with several inputs, all of which will be covered shortly. Next, we individually define the layers within our network, starting with our embedding layer:

```
self.embedding = nn.Embedding(vocab_size, embedding_dim,
padding_idx = pad_idx)
```

The embedding layer will consist of embeddings for each possible word in our vocabulary, so the size of the layer is the length of our vocabulary and the length of our embedding vectors. We are using the 200-dimensional GLoVe vectors, so the length will be 200 in this instance. We must also pass the padding index, which is the index of our embedding layer that's used to get the embedding to pad our sentences so that they are all the same length. We will manually define this embedding later on when we initialize our model.

3. Next, we define the actual convolutional layers within our network:

```
self.convs = nn.ModuleList([
nn.Conv2d(in_channels = 1,
        out_channels = n_filters,
        kernel_size = (fs, embedding_dim))
            for fs in filter_sizes
            ])
```

4. We start by using nn.ModuleList to define a series of convolutional layers. ModuleList takes a list of modules as input and is used when you wish to define a number of separate layers. As we wish to train several different convolutional layers of different sizes on our input data, we can use ModuleList to do so. We could theoretically define each layer separately like so:

```
self.conv_2 = nn.Conv2d(in_channels = 1,
        out_channels = n_filters,
        kernel_size = (2, embedding_dim))

self.conv_3 = nn.Conv2d(in_channels = 1,
        out_channels = n_filters,
        kernel_size = (3, embedding_dim))
```

Here, the filter sizes are 2 and 3, respectively. However, it is more efficient to do this in a single function. Furthermore, our layers will be automatically generated if we pass different filter sizes to the function, rather than having to manually define each layer each time we add a new one.

We also define the out_channels value as the number of filters we wish to train; kernel_size will contain the length of our embeddings. Therefore, we can pass our ModuleList function the lengths of the filters we wish to train and the amount of each and it will automatically generate the convolutional layers. An example of how this convolution layer might look for a given set of variables is as follows:

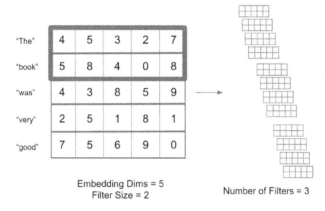

Figure 6.15 – Convolution layer looking for variables

We can see that our ModuleList function adapts to the number of filters and the size of the filters we wish to train. Next, within our CNN initialization, we define the remaining layers, namely the linear layer, which will classify our data, and the dropout layer, which will regularize our network:

```
self.fc = nn.Linear(len(filter_sizes) * n_filters, output_dim)

self.dropout = nn.Dropout(dropout)
```

Note how, in the past, the size of our linear layer has always been 1 as we have only ever needed a single output node to perform binary classification. Since we are now tackling a multi-class classification problem, we wish to make a prediction on each of our potential classes, so our output dimensions are now variable instead of being just 1. When we initialize our network, we will set out output dimensions to be 6 since we are predicting which one of the six classes our sentence is from.

Next, as with all our neural networks, we must define our `forward` pass:

```
def forward(self, text):
emb = self.embedding(text).unsqueeze(1)
conved = [F.relu(c(emb)).squeeze(3) for c in self.convs]
pooled = [F.max_pool1d(c, c.shape[2]).squeeze(2)
          for c in conved]
concat = self.dropout(torch.cat(pooled, dim = 1))
return self.fc(concat)
```

Here, we first pass our input text through our embedding layer to obtain the embeddings for all the words in the sentences. Next, for each of the previously defined convolutional layers that we passed our embedded sentence into, we apply a `relu` activation function and squeeze the results, removing the fourth dimension of the resulting output. This is repeated for all of our defined convolutional layers so that `conved` consists in a list of the outputs for all of our convolutional layers.

For each of these outputs, we apply a pooling function to reduce the dimensionality of our convolutional layer outputs, as described earlier. We then concatenate all the outputs of our pooling layers together and apply a dropout function before passing this to our final fully connected layer, which will make our class predictions. After fully defining our CNN class, we create an instance of the model. We define our hyperparameters and create an instance of the CNN class using them:

```
input_dimensions = len(questions.vocab)
output_dimensions = 6
embedding_dimensions = 200
pad_index = questions.vocab.stoi[questions.pad_token]


number_of_filters = 100
filter_sizes = [2,3,4]
dropout_pc = 0.5


model = CNN(input_dimensions, embedding_dimensions, number_
of_filters, filter_sizes, output_dimensions, dropout_pc, pad_
index)
```

Our input dimensions will always be the length of our vocabulary, while our output dimensions will be the number of classes we wish to predict. Here, we are predicting from six different classes, so our output vector will have a length of 6. Our embedding dimensions are the length of our GLoVe vectors (in this case, 200). The padding index can be manually taken from our vocabulary.

The next three hyperparameters can be manually adjusted, so you may wish to experiment with choosing different values to see how this affects your network's final output. We pass a list of filter sizes so that our model will train convolutional layers using convolutions of size 2, 3, and 4. We will train 100 of these filters for each of these filter sizes, so there will be 300 filters in total. We also define a dropout percentage of 50% for our network to ensure it is sufficiently regularized. This can be raised/lowered if the model seems prone to overfitting or underfitting. A general rule of thumb is to try lowering the dropout rate if the model is underfitting and raising it if the model appears to be overfitting.

After initializing our model, we need to load our weights into our embedding layer. This can be easily done as follows:

```
glove_embeddings = questions.vocab.vectors

model.embedding.weight.data.copy_(glove_embeddings)
```

This results in the following output:

```
tensor([[-0.5928,  1.9557, -0.4180,  ..., -0.1732, -1.0009, -0.7655],
        [-1.1752, -1.5180,  1.7845,  ...,  0.4673, -1.0432,  0.5888],
        [ 0.3911,  0.4019, -0.1505,  ..., -0.0348,  0.0798,  0.5031],
        ...,
        [-0.4949, -0.1262, -1.1698,  ...,  0.5565,  0.5634,  0.5782],
        [ 0.5741, -0.4343, -0.1119,  ...,  0.7629,  0.3831, -0.1570],
        [-0.2597,  0.6716,  0.5353,  ...,  1.6765,  0.3301, -0.1003]])
```

Figure 6.16 – Tensor output after lowering the dropout

Next, we need to define how our model deals with instances when our model accounts for unknown tokens that aren't contained within the embedding layer and how our model will apply padding to our input sentences. Fortunately, the easiest way to account for both of these scenarios is to use a vector consisting of all zeros. We make sure that these zero value tensors are the same length as our embedding vectors (in this instance, 200):

```
unknown_index = questions.vocab.stoi[questions.unk_token]

model.embedding.weight.data[unknown_index] = torch.
zeros(embedding_dimensions)
model.embedding.weight.data[pad_index] = torch.zeros(embedding_
dimensions)
```

Finally, we define our optimizer and criterion (loss) function. Notice how we choose to use cross-entropy loss instead of binary cross-entropy as our classification task is no longer binary. We also use `.to(device)` to train our model using our specified device. This means that our training will be done on a CUDA-enabled GPU if one is available:

```
optimizer = torch.optim.Adam(model.parameters())

criterion = nn.CrossEntropyLoss().to(device)

model = model.to(device)
```

Now that our model's structure has been fully defined, we are ready to start training the model.

Training the CNN

Before we define our training process, we need to calculate a performance metric to illustrate how our model's performance (hopefully!) increases over time. In our binary classification tasks, accuracy was a simple metric we used to measure performance. For our multi-classification task, we will again use accuracy, but the procedure to calculate it is slightly more complex as we must now work out which of the six classes our model predicted and which of the six classes was the correct one.

First, we define a function called `multi_accuracy` to calculate this:

```
def multi_accuracy(preds, y):
    pred = torch.max(preds,1).indices
    correct = (pred == y).float()
    acc = correct.sum() / len(correct)
    return acc
```

Here, for our predictions, our model returns the indices with the highest value for each prediction using the `torch.max` function for all of them. For each of these predictions, if this predicted index is the same as the index of our label, it is treated as a correct prediction. We then count all these correct predictions and divide them by the total number of predictions to get a measure of multi-class accuracy. We can use this function within our training loop to measure accuracy at each epoch.

Next, we define our training function. We initially set our loss and accuracy for the epoch to be 0 and we call `model.train()` to allow the parameters within our model to be updated as we train our model:

```
def train(model, iterator, optimizer, criterion):

    epoch_loss = 0
    epoch_acc = 0

    model.train()
```

Next, we loop through each batch of data within our iterator and perform the training steps. We start by zeroing our gradients to prevent cumulative gradients from being calculated from our previous batch. We then use our model's current state to make predictions from our sentences in the current batch, which we then compare to our labels to calculate loss. Using the accuracy function we defined in the preceding section, we can calculate the accuracy for this given batch. We then backpropagate our loss, updating our weights via gradient descent and step through our optimizer:

```
for batch in iterator:

optimizer.zero_grad()

preds = model(batch.text).squeeze(1)
loss = criterion(preds, batch.label.long())

acc = multi_accuracy(preds, batch.label)

loss.backward()

optimizer.step()
```

Finally, we add the loss and accuracy from this batch to our total loss and accuracy for the whole epoch. After we have looped through all the batches within the epoch, we calculate the total loss and accuracy for the epoch and return it:

```
epoch_loss += loss.item()
epoch_acc += acc.item()


total_epoch_loss = epoch_loss / len(iterator)
total_epoch_accuracy = epoch_acc / len(iterator)


return total_epoch_loss, total_epoch_accuracy
```

Similarly, we can define a function called `eval` that is called on our validation data to calculate our trained model performance on a set of data that our model has not been trained on. While this function is almost identical to the training function we defined previously, there are two key additions we must make:

```
model.eval()


with torch.no_grad():
```

These two steps set our model to evaluation mode, ignore any dropout functions, and make sure that the gradients are not calculated and updated. This is because we want the weights within our model to be frozen while we evaluate performance, as well as to ensure that our model is not trained using our validation data as we want this to be kept separate from the data we used to train the model.

Now, we just need to call our train and evaluate functions in a loop in conjunction with our data iterators in order to train the model. We first define the number of epochs we wish our model to train for. We also define the lowest validation loss our model has achieved so far. This is because we only wish to keep the trained model with the lowest validation loss (that is, the best-performing model). This means that if our model trains for many epochs and begins overfitting, only the best-performing of these models will be kept, meaning there are fewer consequences for picking a high number of epochs.

We initialize our lowest validation loss as infinity to begin with:

```
epochs = 10


lowest_validation_loss = float('inf')
```

Next, we define our training loop, stepping through one epoch at a time. We record the start and end times of our training so that we can calculate how long each step takes. We then simply call our training function on our model using the training data iterator to calculate the training loss and accuracy, updating our model as we do so. We then repeat this process using our evaluation function on our validation iterator to calculate the loss and accuracy on our validation data, without updating our model:

```
for epoch in range(epochs):

    start_time = time.time()

    train_loss, train_acc = train(model, train_iterator,
                            optimizer, criterion)
    valid_loss, valid_acc = evaluate(model, valid_iterator,
                            criterion)

    end_time = time.time()
```

Following this, we determine whether our model, after the current epoch, outperforms our best-performing model so far:

```
if valid_loss < lowest_validation_loss:
    lowest_validation_loss = valid_loss
    torch.save(model.state_dict(), 'cnn_model.pt')
```

If the loss after this epoch is lower than the lowest validation loss so far, we set the validation loss to the new lowest validation loss and save our current model weights.

Finally, we simply print the results after each epoch. If all is working correctly, we should see our training losses fall after every epoch, with our validation losses hopefully following suit:

```
print(f'Epoch: {epoch+1:02} | Epoch Time: {int(end_time -
    start_time)}s')
print(f'\tTrain Loss: {train_loss:.3f} | Train Acc: {train_
    acc*100:.2f}%')
print(f'\t Val. Loss: {valid_loss:.3f} |  Val. Acc: {valid_
    acc*100:.2f}%')
```

This results in the following output:

```
Epoch: 01 | Epoch Time: 7s
        Train Loss: 1.198 | Train Acc: 52.22%
         Val. Loss: 0.859 |  Val. Acc: 69.83%
Epoch: 02 | Epoch Time: 7s
        Train Loss: 0.750 | Train Acc: 73.97%
         Val. Loss: 0.667 |  Val. Acc: 76.32%
Epoch: 03 | Epoch Time: 8s
        Train Loss: 0.515 | Train Acc: 83.02%
         Val. Loss: 0.555 |  Val. Acc: 79.85%
```

Figure 6.17 – Testing the model

Thankfully, we see that this does appear to be the case. Both the training and validation loss fall after every epoch and the accuracy rises, showing that our model is indeed learning! After many training epochs, we can take our best model and use it to make predictions.

Making predictions using the trained CNN

Fortunately, using our fully trained model to make predictions is a relatively simple task. We first load our best model using the `load_state_dict` function:

```
model.load_state_dict(torch.load('cnn_model.pt'))
```

Our model structure has already been defined, so we simply load the weights from the file we saved earlier. If this has worked correctly, you will see the following output:

```
<All keys matched successfully>
```

Figure 6.18 – Prediction output

Next, we define a function that will take a sentence as input, preprocess it, pass it to our model, and return a prediction:

```
def predict_class(model, sentence, min_len = 5):

    tokenized = [tok.text for tok in nlp.tokenizer(sentence)]
    if len(tokenized) < min_len:
        tokenized += ['<pad>'] * (min_len - len(tokenized))
    indexed = [questions.vocab.stoi[t] for t in tokenized]
    tensor = torch.LongTensor(indexed).to(device)
    tensor = tensor.unsqueeze(0)
```

We first pass our input sentence into our tokenizer to get a list of tokens. We then add padding to this sentence if it is below the minimum sentence length. We then use our vocabulary to obtain the index of all these individual tokens before finally creating a tensor consisting of a vector of these indexes. We pass this to our GPU if it is available and then unsqueeze the output as our model expects a three-dimensional tensor input instead of a single vector.

Next, we make our predictions:

```
model.eval()
prediction = torch.max(model(tensor),1).indices.item()
pred_index = labels.vocab.itos[prediction]
    return pred_index
```

We first set our model to evaluation mode (as we did with our evaluation step) so that the gradients of our model are not calculated and the weights are not adjusted. We then pass our sentence tensor into our model and obtain a prediction vector of length 6, consisting of the individual predictions for each of the six classes. We then take the index of the maximum prediction value and use this within our label index to return the name of the predicted class.

In order to make predictions, we simply call the predict_class function on any given sentence. Let's use the following code:

```
pred_class = predict_class(model, "How many roads must a man
                           walk down?")
print('Predicted class is: ' + str(pred_class))
```

This returns the following prediction:

Predicted class is: NUM

Figure 6.19 – Prediction value

This prediction is correct! Our input question contains How many, suggesting that the answer to this question is a numeric value. This is exactly what our model predicts too! You can continue to validate the model on any other questions you might wish to test, hopefully with similarly positive results. Congratulations – you have now successfully trained a multi-class CNN that can define the category of any given question.

Summary

In this chapter, we have shown how CNNs can be used to learn from NLP data and how to train one from scratch using PyTorch. While the deep learning methodology is very different to the methodology used within RNNs, conceptually, CNNs use the motivation behind n-gram language models in an algorithmic fashion in order to extract implicit information about words in a sentence from the context of its neighboring words. Now that we have mastered both RNNs and CNNs, we can begin to expand on these techniques in order to construct even more advanced models.

In the next chapter, we will learn how to build models that utilize elements of both convolutional and recurrent neural networks and use them on sequences to perform even more advanced functions, such as text translation. These are known as sequence-to-sequence networks.

7
Text Translation Using Sequence-to-Sequence Neural Networks

In the previous two chapters, we used neural networks to classify text and perform sentiment analysis. Both tasks involve taking an NLP input and predicting some value. In the case of our sentiment analysis, this was a number between 0 and 1 representing the sentiment of our sentence. In the case of our sentence classification model, our output was a multi-class prediction, of which there were several categories our sentence belonged to. But what if we wish to make not just a single prediction, but predict a whole sentence? In this chapter, we will build a sequence-to-sequence model that takes a sentence in one language as input and outputs the translation of this sentence in another language.

We have already explored several types of neural network architecture used for NLP learning, namely recurrent neural networks in *Chapter 5, Recurrent Neural Networks and Sentiment Analysis*, and convolutional neural networks in *Chapter 6, Text Classification Using CNNs*. In this chapter, we will again be using these familiar RNNs, but instead of just building a simple RNN model, we will use RNNs as part of a larger, more complex model in order to perform sequence-to-sequence translation. By using the underpinnings of RNNs that we learned about in the previous chapters, we can show how these concepts can be extended in order to create a variety of models that can be fit for purpose.

In this chapter, we will cover the following topics:

- Theory of sequence-to-sequence models

- Building a sequence-to-sequence neural network for text translation

- Next steps

Technical requirements

All the code for this chapter can be found at `https://github.com/ PacktPublishing/Hands-On-Natural-Language-Processing-with-PyTorch-1.x`.

Theory of sequence-to-sequence models

Sequence-to-sequence models are very similar to the conventional neural network structures we have seen so far. The main difference is that for a model's output, we expect another sequence, rather than a binary or multi-class prediction. This is particularly useful in tasks such as translation, where we may wish to convert a whole sentence into another language.

In the following example, we can see that our English-to-Spanish translation maps word to word:

Figure 7.1 – English to Spanish translation

The first word in our input sentence maps nicely to the first word in our output sentence. If this were the case for all languages, we could simply pass each word in our sentence one by one through our trained model to get an output sentence, and there would be no need for any sequence-to-sequence modeling, as shown here:

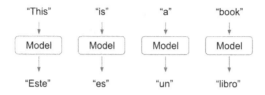

Figure 7.2 – English-to-Spanish translation of words

However, we know from our experience with NLP that language is not as simple as this! Single words in one language may map to multiple words in other languages, and the order in which these words occur in a grammatically correct sentence may not be the same. Therefore, we need a model that can capture the context of a whole sentence and output a correct translation, not a model that aims to directly translate individual words. This is where sequence-to-sequence modeling becomes essential, as seen here:

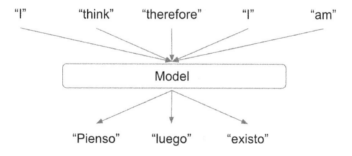

Figure 7.3 – Sequence-to-sequence modeling for translation

To train a sequence-to-sequence model that captures the context of the input sentence and translates this into an output sentence, we will essentially train two smaller models that allow us to do this:

- An **encoder** model, which captures the context of our sentence and outputs it as a single context vector

- A **decoder**, which takes the context vector representation of our original sentence and translates this into a different language

So, in reality, our full sequence-to-sequence translation model will actually look something like this:

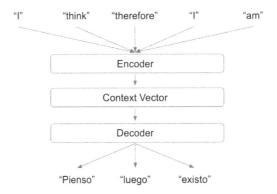

Figure 7.4 – Full sequence-to-sequence model

By splitting our models into individual encoder and decoder elements, we are effectively modularizing our models. This means that if we wish to train multiple models to translate from English into different languages, we do not need to retrain the whole model each time. We only need to train multiple different decoders to transform our context vector into our output sentences. Then, when making predictions, we can simply swap out the decoder that we wish to use for our translation:

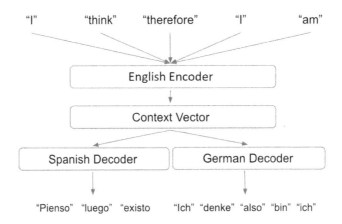

Figure 7.5 – Detailed model layout

Next, we will examine the encoder and decoder components of the sequence-to-sequence model.

Encoders

The purpose of the encoder element of our sequence-to-sequence model is to be able to fully capture the context of our input sentence and represent it as a vector. We can do this by using RNNs or, more specifically, LSTMs. As you may recall from our previous chapters, RNNs take a sequential input and maintain a hidden state throughout this sequence. Each new word in the sequence updates the hidden state. Then, at the end of the sequence, we can use the model's final hidden state as our input into our next layer.

In the case of our encoder, the hidden state represents the context vector representation of our whole sentence, meaning we can use the hidden state output of our RNN to represent the entirety of the input sentence:

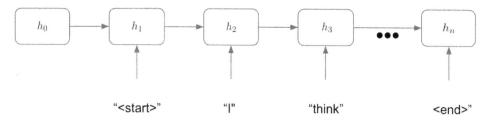

Figure 7.6 – Examining the encoder

We use our final hidden state, h_n, as our context vector, which we will then decode using a trained decoder. It is also worth observing that in the context of our sequence-to-sequence models, we append "start" and "end" tokens to the beginning and end of our input sentence, respectively. This is because our inputs and outputs do not have a finite length and our model needs to be able to learn when a sentence should end. Our input sentence will always end with an "end" token, which signals to the encoder that the hidden state, at this point, will be used as the final context vector representation for this input sentence. Similarly, in the decoder step, we will see that our decoder will keep generating words until it predicts an "end" token. This allows our decoder to generate actual output sentences, as opposed to a sequence of tokens of infinite length.

Next, we will look at how the decoder takes this context vector and learns to translate it into an output sentence.

Decoders

Our decoder takes the final hidden state from our encoder layer and decodes this into a sentence in another language. Our decoder is an RNN, similar to that of our encoder, but while our encoder updates its hidden state given its current hidden state and the current word in the sentence, our decoder updates its hidden state and outputs a token at each iteration, given the current hidden state and the previous predicted word in the sentence. This can be seen in the following diagram:

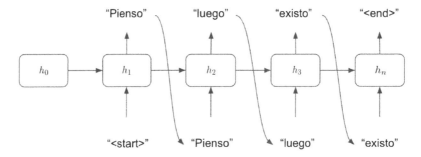

Figure 7.7 – Examining the decoder

First, our model takes the context vector as the final hidden state from our encoder step, *h0*. Our model then aims to predict the next word in the sentence, given the current hidden state, and then the previous word in the sentence. We know our sentence must begin with a "start" token so, at our first step, our model tries to predict the first word in the sentence given the previous hidden state, *h0*, and the previous word in the sentence (in this instance, the "start" token). Our model makes a prediction ("pienso") and then updates the hidden state to reflect the new state of the model, *h1*. Then, at the next step, our model uses the new hidden state and the last predicted word to predict the next word in the sentence. This continues until the model predicts the "end" token, at which point our model stops generating output words.

The intuition behind this model is in line with what we have learned about language representations thus far. Words in any given sentence are dependent on the words that come before it. So, to predict any given word in a sentence without considering the words that have been predicted before it, this would not make sense as words in any given sentence are not independent from one another.

We learn our model parameters as we have done previously: by making a forward pass, calculating the loss of our target sentence against the predicted sentence, and backpropagating this loss through the network, updating the parameters as we go. However, learning using this process can be very slow because, to begin with, our model will have very little predictive power. Since our predictions for the words in our target sentence are not independent of one another, if we predict the first word in our target sentence incorrectly, subsequent words in our output sentence are also unlikely to be correct. To help with this process, we can use a technique known as **teacher forcing**.

Using teacher forcing

As our model does not make good predictions initially, we will find that any initial errors are multiplied exponentially. If our first predicted word in the sentence is incorrect, then the rest of the sentence will likely be incorrect as well. This is because the predictions our model makes are dependent on the previous predictions it makes. This means that any losses our model has can be multiplied exponentially. Due to this, we may face the exploding gradient problem, making it very difficult for our model to learn anything:

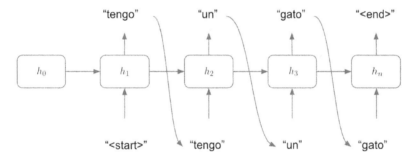

Figure 7.8 – Using teacher forcing

However, by using **teacher forcing**, we train our model using the correct previous target word so that one wrong prediction does not inhibit our model's ability to learn from the correct predictions. This means that if our model makes an incorrect prediction at one point in the sentence, it can still make correct predictions using subsequent words. While our model will still have incorrectly predicted words and will have losses by which we can update our gradients, now, we do not suffer from exploding gradients, and our model will learn much more quickly:

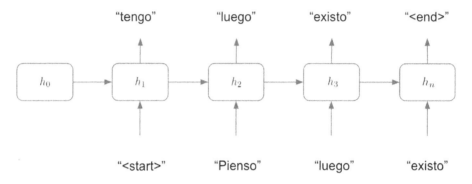

Figure 7.9 – Updating for losses

You can consider teacher forcing as a way of helping our model learn independently of its previous predictions at each time step. This is so the losses that are incurred by a mis-prediction at an early time step are not carried over to later time steps.

By combining the encoder and decoder steps and applying teacher forcing to help our model learn, we can build a sequence-to-sequence model that will allow us to translate sequences of one language into another. In the next section, we will illustrate how we can build this from scratch using PyTorch.

Building a sequence-to-sequence model for text translation

In order to build our sequence-to-sequence model for translation, we will implement the encoder/decoder framework we outlined previously. This will show how the two halves of our model can be utilized together in order to capture a representation of our data using the encoder and then translate this representation into another language using our decoder. In order to do this, we need to obtain our data.

Preparing the data

By now, we know enough about machine learning to know that for a task like this, we will need a set of training data with corresponding labels. In this case, we will need **sentences in one language with the corresponding translations in another language**. Fortunately, the Torchtext library that we used in the previous chapter contains a dataset that will allow us to get this.

The Multi30k dataset in Torchtext consists of approximately 30,000 sentences with corresponding translations in multiple languages. For this translation task, our input sentences will be in English and our output sentences will be in German. Our fully trained model will, therefore, allow us to **translate English sentences into German**.

We will start by extracting our data and preprocessing it. We will once again use spacy, which contains a built-in dictionary of vocabulary that we can use to tokenize our data:

1. We start by loading our spacy tokenizers into Python. We will need to do this once for each language we are using since we will be building two entirely separate vocabularies for this task:

```
spacy_german = spacy.load('de')
spacy_english = spacy.load('en')
```

> **Important note**
>
> You may have to install the German vocabulary from the command line by doing the following (we installed the English vocabulary in the previous chapter):
>
> **python3 -m spacy download de**

2. Next, we create a function for each of our languages to tokenize our sentences. Note that our tokenizer for our input English sentence reverses the order of the tokens:

```
def tokenize_german(text):
    return [token.text for token in spacy_german.
            tokenizer(text)]

def tokenize_english(text):
    return [token.text for token in spacy_english.
            tokenizer(text)][::-1]
```

While reversing the order of our input sentence is not compulsory, it has been shown to improve the model's ability to learn. If our model consists of two RNNs joined together, we can show that the information flow within our model is improved when reversing the input sentence. For example, let's take a basic input sentence in English but not reverse it, as follows:

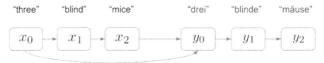

Figure 7.10 – Reversing the input words

Here, we can see that in order to predict the first output word, $y0$, correctly, our first English word from $x0$ must travel through three RNN layers before the prediction is made. In terms of learning, this means that our gradients must be backpropagated through three RNN layers, while maintaining the flow of information through the network. Now, let's compare this to a situation where we reverse our input sentence:

Figure 7.11 – Reversing the input sentence

We can now see that the distance between the true first word in our input sentence and the corresponding word in the output sentence is just one RNN layer. This means that the gradients only need to be backpropagated to one layer, meaning the flow of information and the ability to learn is much greater for our network compared to when the distance between these two words was three layers.

If we were to calculate the total distances between the input words and their output counterparts for the reversed and non-reversed variants, we would see that they are the same. However, we have seen previously that the most important word in our output sentence is the first one. This is because the words in our output sentences are dependent on the words that come before them. If we were to predict the first word in the output sentence incorrectly, then chances are the rest of the words in our sentences would be predicted incorrectly too. However, by predicting the first word correctly, we maximize our chances of predicting the whole sentence correctly. Therefore, by minimizing the distance between the first word in our output sentence and its input counterpart, we can increase our model's ability to learn this relationship. This increases the chances of this prediction being correct, thus maximizing the chances of our entire output sentence being predicted correctly.

3. With our tokenizers constructed, we now need to define the fields for our tokenization. Notice here how we append start and end tokens to our sequences so that our model knows when to begin and end the sequence's input and output. We also convert all our input sentences into lowercase for the sake of simplicity:

```
SOURCE = Field(tokenize = tokenize_english,
               init_token = '<sos>',
               eos_token = '<eos>',
               lower = True)

TARGET = Field(tokenize = tokenize_german,
               init_token = '<sos>',
               eos_token = '<eos>',
               lower = True)
```

4. With our fields defined, our tokenization becomes a simple one-liner. The dataset containing 30,000 sentences has built-in training, validation, and test sets that we can use for our model:

```
train_data, valid_data, test_data = Multi30k.splits(exts
= ('.en', '.de'), fields = (SOURCE, TARGET))
```

5. We can examine individual sentences using the `examples` property of our dataset objects. Here, we can see that the source (`src`) property contains our reversed input sentence in English and that our target (`trg`) contains our non-reversed output sentence in German:

```
print(train_data.examples[0].src)
print(train_data.examples[0].trg)
```

This gives us the following output:

```
['.', 'bushes', 'many', 'near', 'outside', 'are', 'males',
'white', ',', 'young', 'two']
['zwei', 'junge', 'weiße', 'männer', 'sind', 'im', 'freie
n', 'in', 'der', 'nähe', 'vieler', 'büsche', '.']
```

Figure 7.12 – Training data examples

6. Now, we can examine the size of each of our datasets. Here, we can see that our training dataset consists of 29,000 examples and that each of our validation and test sets consist of 1,014 and 1,000 examples, respectively. In the past, we have used 80%/20% splits for the training and validation data. However, in instances like this, where our input and output fields are very sparse and our training set is of a limited size, it is often beneficial to train on as much data as there is available:

```
print("Training dataset size: " + str(len(train_data.
    examples)))
print("Validation dataset size: " + str(len(valid_data.
    examples)))
print("Test dataset size: " + str(len(test_data.
    examples)))
```

This returns the following output:

```
Training dataset size: 29000
Validation dataset size: 1014
Test dataset size: 1000
```

Figure 7.13 – Data sample lengths

7. Now, we can build our vocabularies and check their size. Our vocabularies should consist of every unique word that was found within our dataset. We can see that our German vocabulary is considerably larger than our English vocabulary. Our vocabularies are significantly smaller than the true size of each vocabulary for each language (every word in the English dictionary). Therefore, since our model will only be able to accurately translate words it has seen before, it is unlikely that our model will be able to generalize well to all possible sentences in the English language. This is why training models like this accurately requires extremely large NLP datasets (such as those Google has access to):

```
SOURCE.build_vocab(train_data, min_freq = 2)
TARGET.build_vocab(train_data, min_freq = 2)

print("English (Source) Vocabulary Size: " +
    str(len(SOURCE.vocab)))
print("German (Target) Vocabulary Size: " +
    str(len(TARGET.vocab)))
```

This gives the following output:

```
English (Source) Vocabulary Size: 5893
German (Target) Vocabulary Size: 7855
```

Figure 7.14 – Vocabulary size of the dataset

8. Finally, we can create our data iterators from our datasets. As we did previously, we specify the usage of a CUDA-enabled GPU (if it is available on our system) and specify our batch size:

```
device = torch.device('cuda' if torch.cuda.is_available()
                    else 'cpu')


batch_size = 32


train_iterator, valid_iterator, test_iterator =
BucketIterator.splits(
    (train_data, valid_data, test_data),
    batch_size = batch_size,
    device = device)
```

Now that our data has been preprocessed, we can start building the model itself.

Building the encoder

Now, we are ready to start building our encoder:

1. First, we begin by initializing our model by inheriting from our nn.Module class, as we've done with all our previous models. We initialize with a couple of parameters, which we will define later, as well as the number of dimensions in the hidden layers within our LSTM layers and the number of LSTM layers:

```
class Encoder(nn.Module):
    def __init__(self, input_dims, emb_dims, hid_dims,
    n_layers, dropout):
        super().__init__()
        self.hid_dims = hid_dims
        self.n_layers = n_layers
```

2. Next, we define our embedding layer within our encoder, which is the length of the number of input dimensions and the depth of the number of embedding dimensions:

```
self.embedding = nn.Embedding(input_dims, emb_dims)
```

3. Next, we define our actual LSTM layer. This takes our embedded sentences from the embedding layer, maintains a hidden state of a defined length, and consists of a number of layers (which we will define later as 2). We also implement dropout to apply regularization to our network:

```
self.rnn = nn.LSTM(emb_dims, hid_dims, n_layers, dropout
                  = dropout)
self.dropout = nn.Dropout(dropout)
```

4. Then, we define the forward pass within our encoder. We apply the embeddings to our input sentences and apply dropout. Then, we pass these embeddings through our LSTM layer, which outputs our final hidden state. This will be used by our decoder to form our translated sentence:

```
def forward(self, src):
    embedded = self.dropout(self.embedding(src))
    outputs, (h, cell) = self.rnn(embedded)
    return h, cell
```

Our encoders will consist of two LSTM layers, which means that our output will output two hidden states. This also means that our full LSTM layer, along with our encoder, will look something like this, with our model outputting two hidden states:

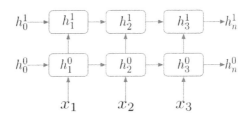

Figure 7.15 – LSTM model with an encoder

Now that we have built our encoder, let's start building our decoder.

Building the decoder

Our decoder will take the final hidden states from our encoder's LSTM layer and translate them into an output sentence in another language. We start by initializing our decoder in almost exactly the same way as we did for the encoder. The only difference here is that we also add a fully connected linear layer. This layer will use the final hidden states from our LSTM in order to make predictions regarding the correct word in the sentence:

```python
class Decoder(nn.Module):
    def __init__(self, output_dims, emb_dims, hid_dims,
    n_layers, dropout):
        super().__init__()

        self.output_dims = output_dims
        self.hid_dims = hid_dims
        self.n_layers = n_layers

        self.embedding = nn.Embedding(output_dims, emb_dims)

        self.rnn = nn.LSTM(emb_dims, hid_dims, n_layers,
                           dropout = dropout)

        self.fc_out = nn.Linear(hid_dims, output_dims)

        self.dropout = nn.Dropout(dropout)
```

Our forward pass is incredibly similar to that of our encoder, except with the addition of two key steps. We first unsqueeze our input from the previous layer so that it's the correct size for entry into our embedding layer. We also add a fully connected layer, which takes the output hidden layer of our RNN layers and uses it to make a prediction regarding the next word in the sequence:

```python
def forward(self, input, h, cell):

    input = input.unsqueeze(0)

    embedded = self.dropout(self.embedding(input))

    output, (h, cell) = self.rnn(embedded, (h, cell))
```

```
pred = self.fc_out(output.squeeze(0))

return pred, h, cell
```

Again, similar to our encoder, we use a two-layer LSTM layer within our decoder. We take our final hidden state from our encoders and use these to generate the first word in our sequence, Y_1. We then update our hidden state and use this and Y_1 to generate our next word, Y_2, repeating this process until our model generates an end token. Our decoder looks something like this:

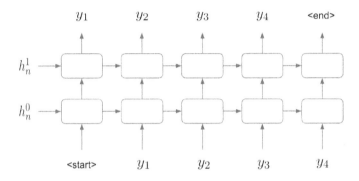

Figure 7.16 – LSTM model with a decoder

Here, we can see that defining the encoders and decoders individually is not particularly complicated. However, when we combine these steps into one larger sequence-to-sequence model, things begin to get interesting:

Constructing the full sequence-to-sequence model

We must now stitch the two halves of our model together to produce the full sequence-to-sequence model:

1. We start by creating a new sequence-to-sequence class. This will allow us to pass our encoder and decoder to it as arguments:

```
class Seq2Seq(nn.Module):
    def __init__(self, encoder, decoder, device):
        super().__init__()

        self.encoder = encoder
        self.decoder = decoder
        self.device = device
```

2. Next, we create the `forward` method within our `Seq2Seq` class. This is arguably the most complicated part of the model. We combine our encoder with our decoder and use teacher forcing to help our model learn. We start by creating a tensor in which we still store our predictions. We initialize this as a tensor full of zeroes, but we still update this with our predictions as we make them. The shape of our tensor of zeroes will be the length of our target sentence, the width of our batch size, and the depth of our target (German) vocabulary size:

```
def forward(self, src, trg, teacher_forcing_rate = 0.5):
    batch_size = trg.shape[1]
    target_length = trg.shape[0]
    target_vocab_size = self.decoder.output_dims

    outputs = torch.zeros(target_length, batch_size,
                     target_vocab_size).to(self.device)
```

3. Next, we feed our input sentence into our encoder to get the output hidden states:

```
h, cell = self.encoder(src)
```

4. Then, we must loop through our decoder model to generate an output prediction for each step in our output sequence. The first element of our output sequence will always be the `<start>` token. Our target sequences already contain this as the first element, so we just set our initial input equal to this by taking the first element of the list:

```
input = trg[0,:]
```

5. Next, we loop through and make our predictions. We pass our hidden states (from the output of our encoder) to our decoder, along with our initial input (which is just the `<start>` token). This returns a prediction for all the words in our sequence. However, we are only interested in the word within our current step; that is, the next word in the sequence. Note how we start our loop from 1 instead of 0, so our first prediction is the second word in the sequence (as the first word that's predicted will always be the start token).

6. This output consists of a vector of the target vocabulary's length, with a prediction for each word within the vocabulary. We take the `argmax` function to identify the actual word that is predicted by the model.

We then need to select our new input for the next step. We set our teacher forcing ratio to 50%, which means that 50% of the time, we will use the prediction we just made as our next input into our decoder and that the other 50% of the time, we will take the true target. As we discussed previously, this helps our model learn much more rapidly than relying on just the model's predictions.

We then continue this loop until we have a full prediction for each word in the sequence:

```
for t in range(1, target_length):

    output, h, cell = self.decoder(input, h, cell)

    outputs[t] = output

    top = output.argmax(1)

    input = trg[t] if (random.random() < teacher_forcing_
                        rate) else top

return outputs
```

7. Finally, we create an instance of our Seq2Seq model that's ready to be trained. We initialize an encoder and a decoder with a selection of hyperparameters, all of which can be changed to slightly alter the model:

```
input_dimensions = len(SOURCE.vocab)
output_dimensions = len(TARGET.vocab)
encoder_embedding_dimensions = 256
decoder_embedding_dimensions = 256
hidden_layer_dimensions = 512
number_of_layers = 2
encoder_dropout = 0.5
decoder_dropout = 0.5
```

8. We then pass our encoder and decoder to our `Seq2Seq` model in order to create the complete model:

```
encod = Encoder(input_dimensions,\
                encoder_embedding_dimensions,\
                hidden_layer_dimensions,\
                number_of_layers, encoder_dropout)
decod = Decoder(output_dimensions,\
                decoder_embedding_dimensions,\
                hidden_layer_dimensions,\
                number_of_layers, decoder_dropout)

model = Seq2Seq(encod, decod, device).to(device)
```

Try experimenting with different parameters here and see how it affects the performance of the model. For instance, having a larger number of dimensions in your hidden layers may cause the model to train slower, although the overall final performance of the model may be better. Alternatively, the model may overfit. Often, it is a matter of experimenting to find the best-performing model.

After fully defining our Seq2Seq model, we are now ready to begin training it.

Training the model

Our model will begin initialized with weights of 0 across all parts of the model. While the model should theoretically be able to learn with no (zero) weights, it has been shown that initializing with random weights can help the model learn faster. Let's get started:

1. Here, we will initialize our model with the weights of random samples from a normal distribution, with the values being between -0.1 and 0.1:

```
def initialize_weights(m):
    for name, param in m.named_parameters():
        nn.init.uniform_(param.data, -0.1, 0.1)

model.apply(initialize_weights)
```

2. Next, as with all our other models, we define our optimizer and loss functions.
    We're using cross-entropy loss as we are performing multi-class classification
    (as opposed to binary cross-entropy loss for a binary classification):

```
optimizer = optim.Adam(model.parameters())


criterion = nn.CrossEntropyLoss(ignore_index = TARGET.
                vocab.stoi[TARGET.pad_token])
```

3. Next, we define the training process within a function called `train()`. First, we set
    our model to train mode and set the epoch loss to 0:

```
def train(model, iterator, optimizer, criterion, clip):
    model.train()
    epoch_loss = 0
```

4. We then loop through each batch within our training iterator and extract the
    sentence to be translated (`src`) and the correct translation of this sentence (`trg`).
    We then zero our gradients (to prevent gradient accumulation) and calculate the
    output of our model by passing our model function our inputs and outputs:

```
for i, batch in enumerate(iterator):
src = batch.src
trg = batch.trg
optimizer.zero_grad()
output = model(src, trg)
```

5. Next, we need to calculate the loss of our model's prediction by comparing our predicted output to the true, correct translated sentence. We reshape our output data and our target data using the shape and view functions in order to create two tensors that can be compared to calculate the loss. We calculate the `loss` criterion between our output and `trg` tensors and then backpropagate this loss through the network:

```
output_dims = output.shape[-1]
output = output[1:].view(-1, output_dims)
trg = trg[1:].view(-1)


loss = criterion(output, trg)


loss.backward()
```

6. We then implement gradient clipping to prevent exploding gradients within our model, step our optimizer in order to perform the necessary parameter updates via gradient descent, and finally add the loss of the batch to the epoch loss. This whole process is repeated for all the batches within a single training epoch, whereby the final averaged loss per batch is returned:

```
torch.nn.utils.clip_grad_norm_(model.parameters(), clip)


optimizer.step()


epoch_loss += loss.item()


return epoch_loss / len(iterator)
```

7. After, we create a similar function called `evaluate()`. This function will calculate the loss of our validation data across the network in order to evaluate how our model performs when translating data it hasn't seen before. This function is almost identical to our `train()` function, with the exception of the fact that we switch to evaluation mode:

```
model.eval()
```

8. Since we don't perform any updates for our weights, we need to make sure to implement no_grad mode:

```
with torch.no_grad():
```

9. The only other difference is that we need to make sure we turn off teacher forcing when in evaluation mode. We wish to assess our model's performance on unseen data, and enabling teacher forcing would use our correct (target) data to help our model make better predictions. We want our model to be able to make perfect, unaided predictions:

```
output = model(src, trg, 0)
```

10. Finally, we need to create a training loop, within which our train() and evaluate() functions are called. We begin by defining how many epochs we wish to train for and our maximum gradient (for use with gradient clipping). We also set our lowest validation loss to infinity. This will be used later to select our best-performing model:

```
epochs = 10
grad_clip = 1

lowest_validation_loss = float('inf')
```

11. We then loop through each of our epochs and within each epoch, calculate our training and validation loss using our train() and evaluate() functions. We also time how long this takes by calling time.time() before and after the training process:

```
for epoch in range(epochs):

    start_time = time.time()

    train_loss = train(model, train_iterator, optimizer,
                       criterion, grad_clip)
    valid_loss = evaluate(model, valid_iterator,
                         criterion)

    end_time = time.time()
```

12. Next, for each epoch, we determine whether the model we just trained is the best-performing model we have seen thus far. If our model performs the best on our validation data (if the validation loss is the lowest we have seen so far), we save our model:

```
if valid_loss < lowest_validation_loss:
lowest_validation_loss = valid_loss
torch.save(model.state_dict(), 'seq2seq.pt')
```

13. Finally, we simply print our output:

```
print(f'Epoch: {epoch+1:02} | Time: {np.round(end_time-
start_time,0)}s')
print(f'\tTrain Loss: {train_loss:.4f}')
print(f'\t Val. Loss: {valid_loss:.4f}')
```

If our training is working correctly, we should see the training loss decrease over time, like so:

```
Epoch: 01 | Time: 2335.0s
        Train Loss: 4.7367
         Val. Loss: 4.6161
Epoch: 02 | Time: 2294.0s
        Train Loss: 4.0204
         Val. Loss: 4.2858
Epoch: 03 | Time: 2301.0s
        Train Loss: 3.6623
         Val. Loss: 4.0794
Epoch: 04 | Time: 2315.0s
        Train Loss: 3.4008
         Val. Loss: 3.9103
Epoch: 05 | Time: 2305.0s
        Train Loss: 3.1771
         Val. Loss: 3.8092
```

Figure 7.17 – Training the model

Here, we can see that both our training and validation loss appear to be falling over time. We can continue to train our model for a number of epochs, ideally until the validation loss reaches its lowest possible value. Now, we can evaluate our best-performing model to see how well it performs when making actual translations.

Evaluating the model

In order to evaluate our model, we will take our test set of data and run our English sentences through our model to obtain a prediction of the translation in German. We will then be able to compare this to the true prediction in order to see if our model is making accurate predictions. Let's get started!

1. We start by creating a `translate()` function. This is functionally identical to the `evaluate()` function we created to calculate the loss over our validation set. However, this time, we are not concerned with the loss of our model, but rather the predicted output. We pass the model our source and target sentences and also make sure we turn teacher forcing off so that our model does not use these to make predictions. We then take our model's predictions and use an `argmax` function to determine the index of the word that our model predicted for each word in our predicted output sentence:

    ```
    output = model(src, trg, 0)
    preds = torch.tensor([[torch.argmax(x).item()] for x
            in output])
    ```

2. Then, we can use this index to obtain the actual predicted word from our German vocabulary. Finally, we compare the English input to our model that contains the correct German sentence and the predicted German sentence. Note that here, we use `[1:-1]` to drop the start and end tokens from our predictions and we reverse the order of our English input (since the input sentences were reversed before they were fed into the model):

    ```
    print('English Input: ' + str([SOURCE.vocab.itos[x] for x
            in src][1:-1][::-1]))
    print('Correct German Output: ' + str([TARGET.vocab.
            itos[x] for x in trg][1:-1]))
    print('Predicted German Output: ' + str([TARGET.vocab.
            itos[x] for x in preds][1:-1]))
    ```

By doing this, we can compare our predicted output with the correct output to assess if our model is able to make accurate predictions. We can see from our model's predictions that our model is able to translate English sentences into German, albeit far from perfectly. Some of our model's predictions are exactly the same as the target data, showing that our model translated these sentences perfectly:

```
English Input: ['a', 'woman', 'is', 'playing', 'volleyball', '.']
Correct German Output: ['eine', 'frau', 'spielt', 'volleyball', '.']
Predicted German Output: ['eine', 'frau', 'spielt', 'volleyball', '.']
```

Figure 7.18 – Translation output part one

In other instances, our model is off by a single word. In this case, our model predicts the word hüten instead of mützen; however, hüten is actually an acceptable translation of mützen, though the words may not be semantically identical:

```
English Input: ['two', 'men', 'wearing', 'hats', '.']
Correct German Output: ['zwei', 'männer', 'mit', 'mützen', '.']
Predicted German Output: ['zwei', 'männer', 'mit', 'hüten', '.']
```

Figure 7.19 – Translation output part two

We can also see examples that seem to have been mistranslated. In the following example, the English equivalent of the German sentence that we predicted is "A woman climbs through one", which is not equivalent to "Young woman climbing rock face". However, the model has still managed to translate key elements of the English sentence (woman and climbing):

```
English Input: ['young', 'woman', 'climbing', 'rock', 'face']
Correct German Output: ['junge', 'frau', 'klettert', 'auf', 'felswand']
Predicted German Output: ['eine', 'frau', 'klettert', 'durch', 'einen']
```

Figure 7.20 – Translation output part three

Here, we can see that although our model clearly makes a decent attempt at translating English into German, it is far from perfect and makes several mistakes. It certainly would not be able to fool a native German speaker! Next, we will discuss a couple of ways we could improve our sequence-to-sequence translation model.

Next steps

While we have shown our sequence-to-sequence model to be effective at performing language translation, the model we trained from scratch is not a perfect translator by any means. This is, in part, due to the relatively small size of our training data. We trained our model on a set of 30,000 English/German sentences. While this might seem very large, in order to train a perfect model, we would require a training set that's several orders of magnitude larger.

In theory, we would require several examples of each word in the entire English and German languages for our model to truly understand its context and meaning. For context, the 30,000 English sentences in our training set consisted of just 6,000 unique words. The average vocabulary of an English speaker is said to be between 20,000 and 30,000 words, which gives us an idea of just how many examples sentences we would need to train a model that performs perfectly. This is probably why the most accurate translation tools are owned by companies with access to vast amounts of language data (such as Google).

Summary

In this chapter, we covered how to build sequence-to-sequence models from scratch. We learned how to code up our encoder and decoder components individually and how to integrate them into a single model that is able to translate sentences from one language into another.

Although our sequence-to-sequence model, which consists of an encoder and a decoder, is useful for sequence translation, it is no longer state-of-the-art. In the last few years, combining sequence-to-sequence models with attention models has been done to achieve state-of-the-art performance.

In the next chapter, we will discuss how attention networks can be used in the context of sequence-to-sequence learning and show how we can use both techniques to build a chat bot.

8
Building a Chatbot Using Attention-Based Neural Networks

If you have ever watched any futuristic sci-fi movies, chances are you will have seen a human talk to a robot. Machine-based intelligence has been a long-standing feature in works of fiction; however, thanks to recent advances in NLP and deep learning, conversations with a computer are no longer a fantasy. While we may be many years away from true intelligence, where computers are able to understand the meaning of language in the same way that humans do, machines are at least capable of holding a basic conversation and giving a rudimentary impression of intelligence.

In the previous chapter, we looked at how to construct sequence-to-sequence models to translate sentences from one language into another. A conversational chatbot that is capable of basic interactions works in much the same way. When we talk to a chatbot, our sentence becomes the input to the model. The output is whatever the chatbot chooses to reply with. Therefore, rather than training our chatbot to learn how to interpret our input sentence, we are teaching it how to respond.

We will expand on our sequence-to-sequence models from the previous chapter, adding something called **attention** to our models. This improvement to the sequence-to-sequence models means that our model learns where in the input sentence to look to obtain the information it needs, rather than using the whole input sentence decision. This improvement allows us to create much more efficient sequence-to-sequence models with state-of-the-art performance.

In this chapter, we will look at the following topics:

- The theory of attention within neural networks
- Implementing attention within a neural network to construct a chatbot

Technical requirements

All of the code for this chapter can be found at `https://github.com/PacktPublishing/Hands-On-Natural-Language-Processing-with-PyTorch-1.x`.

The theory of attention within neural networks

In the previous chapter, in our sequence-to-sequence model for sentence translation (with no attention implemented), we used both encoders and decoders. The encoder obtained a hidden state from the input sentence, which was a representation of our sentence. The decoder then used this hidden state to perform the translation steps. A basic graphical illustration of this is as follows:

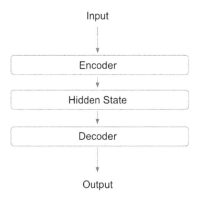

Figure 8.1 – Graphical representation of sequence-to-sequence models

However, decoding over the entirety of the hidden state is not necessarily the most efficient way of using this task. This is because the hidden state represents the entirety of the input sentence; however, in some tasks (such as predicting the next word in a sentence), we do not need to consider the entirety of the input sentence, just the parts that are relevant to the prediction we are trying to make. We can show that by using attention within our sequence-to-sequence neural network. We can teach our model to only look at the relevant parts of the input in order to make its prediction, resulting in a much more efficient and accurate model.

Consider the following example:

I will be traveling to Paris, the capital city of France, on the 2nd of March. My flight leaves from London Heathrow airport and will take approximately one hour.

Let's say that we are training a model to predict the next word in a sentence. We can first input the start of the sentence:

The capital city of France is _____.

We would expect our model to be able to retrieve the word **Paris**, in this case. If we were to use our basic sequence-to-sequence model, we would transform our entire input into a hidden state, which our model would then try to extract the relevant information out of. This includes all the extraneous information about flights. You may notice here that we only need to look at a small part of our input sentence in order to identify the relevant information required to complete our sentence:

*I will be traveling to **Paris, the capital city of France**, on the 2nd of March. My flight leaves from London Heathrow airport and will take approximately one hour.*

Therefore, if we can train our model to only use the relevant information within the input sentence, we can make more accurate and relevant predictions. We can implement **attention** within our networks in order to achieve this.

There are two main types of attention mechanisms that we can implement: local and global attention.

Comparing local and global attention

The two forms of attention that we can implement within our networks are very similar, but with subtle key differences. We will start by looking at local attention.

In **local attention**, our model only looks at a few hidden states from the encoder. For example, if we are performing a sentence translation task and we are calculating the second word in our translation, the model may wish to only look at the hidden states from the encoder related to the second word in the input sentence. This would mean that our model needs to look at the second hidden state from our encoder (h_2) but maybe also the hidden state before it (h_1).

In the following diagram, we can see this in practice:

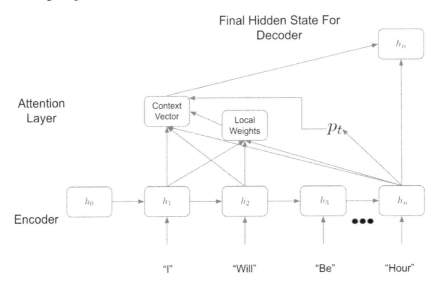

Figure 8.2 – Local attention model

We first start by calculating the aligned position, p_t, from our final hidden state, h_n. This tells us which hidden states we need to be looking at to make our prediction. We then calculate our local weights and apply them to our hidden states in order to determine our context vector. These weights may tell us to pay more attention to the most relevant hidden state (h_2) but less attention to the preceding hidden state (h_1).

We then take our context vector and pass it forward to our decoder in order to make its prediction. In our non-attention based sequence-to-sequence model, we would have only passed our final hidden state, h_n, forward, but we see here that instead, we only consider the relevant hidden states that our model deems necessary to make its prediction.

The **global attention** model works in a very similar way. However, instead of only looking at a few of the hidden states, we want to look at all of our model's hidden states—hence the name global. We can see a graphical illustration of a global attention layer here:

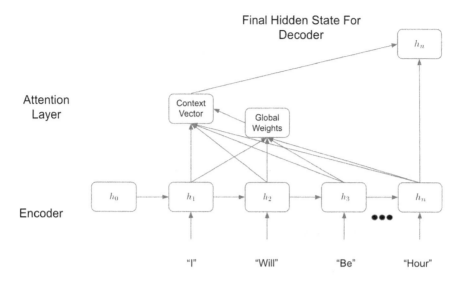

Figure 8.3 – Global attention model

We can see in the preceding diagram that although this appears very similar to our local attention framework, our model is now looking at all the hidden states and calculating the global weights across all of them. This allows our model to look at any given part of the input sentence that it considers relevant, instead of being limited to a local area determined by the local attention methodology. Our model may wish to only look at a small, local area, but this is within the capabilities of the model. An easy way to think of the global attention framework is that it is essentially learning a mask that only allows through hidden states that are relevant to our prediction:

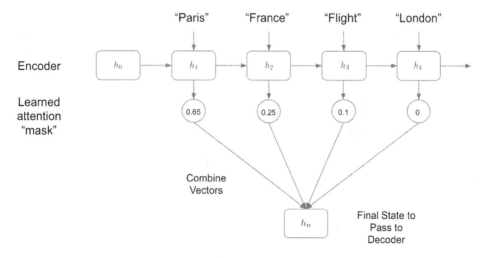

Figure 8.4 – Combined model

We can see in the preceding diagram that by learning which hidden states to pay attention to, our model controls which states are used in the decoding step to determine our predicted output. Once we decide which hidden states to pay attention to, we can combine them using a number of different methods—either by concatenating or taking the weighted dot product.

Building a chatbot using sequence-to-sequence neural networks with attention

The easiest way to illustrate exactly how to implement attention within our neural network is to work through an example. We will now go through the steps required to build a chatbot from scratch using a sequence-to-sequence model with an attention framework applied.

As with all of our other NLP models, our first step is to obtain and process a dataset to use to train our model.

Acquiring our dataset

To train our chatbot, we need a dataset of conversations by which our model can learn how to respond. Our chatbot will take a line of human-entered input and respond to it with a generated sentence. Therefore, an ideal dataset would consist of a number of lines of dialogue with appropriate responses. The perfect dataset for a task such as this would be actual chat logs from conversations between two human users. Unfortunately, this data consists of private information and is very hard to come by within the public domain, so for this task, we will be using a dataset of movie scripts.

Movie scripts consist of conversations between two or more characters. While this data is not naturally in the format we would like it to be in, we can easily transform it into the format that we need. Take, for example, a simple conversation between two characters:

- **Line 1**: Hello Bethan.
- **Line 2**: Hello Tom, how are you?
- **Line 3**: I'm great thanks, what are you doing this evening?
- **Line 4**: I haven't got anything planned.
- **Line 5**: Would you like to come to dinner with me?

Now, we need to transform this into input and output pairs of call and response, where the input is a line in the script (the call) and the expected output is the next line of the script (the response). We can transform a script of *n* lines into *n-1* pairs of input/output:

Input (Call)	Output (Response)
Hello Bethan.	Hello Tom, how are you?
Hello Tom, how are you?	I'm great thanks, what are you doing this evening?
I'm great thanks, what are you doing this evening?	I haven't got anything planned.
I haven't got anything planned.	Would you like to come to dinner with me?

Figure 8.5 – Table of input and output

We can use these input/output pairs to train our network, where the input is a proxy for human input and the output is the response that we would expect from our model.

The first step in building our model is to read this data in and perform all the necessary preprocessing steps.

Processing our dataset

Fortunately, the dataset provided for this example has already been formatted so that each line represents a single input/output pair. We can first read the data in and examine some lines:

```
corpus = "movie_corpus"
corpus_name = "movie_corpus"

datafile = os.path.join(corpus, "formatted_movie_lines.txt")

with open(datafile, 'rb') as file:
    lines = file.readlines()

for line in lines[:3]:
    print(str(line) + '\n')
```

This prints the following result:

```
b"Can we make this quick?  Roxanne Korrine and Andrew Barrett are having
an incredibly horrendous public break- up on the quad.  Again.\tWell, I t
hought we'd start with pronunciation, if that's okay with you.\n"

b"Well, I thought we'd start with pronunciation, if that's okay with yo
u.\tNot the hacking and gagging and spitting part.  Please.\n"

b"Not the hacking and gagging and spitting part.  Please.\tOkay... then h
ow 'bout we try out some French cuisine.  Saturday?  Night?\n"
```

Figure 8.6 – Examining the dataset

You will first notice that our lines are as expected, as the second half of the first line becomes the first half of the next line. We can also note that the call and response halves of each line are separated by a tab delimiter (/t) and that each of our lines is separated by a new line delimiter (/n). We will have to account for this when we process our dataset.

The first step is to create a vocabulary or corpus that contains all the unique words within our dataset.

Creating the vocabulary

In the past, our corpus has comprised of several dictionaries consisting of the unique words in our corpus and lookups between word and indices. However, we can do this in a far more elegant way by creating a vocabulary class that consists of all of the elements required:

1. We start by creating our Vocabulary class. We initialize this class with empty dictionaries—word2index and word2count. We also initialize the index2word dictionary with placeholders for our padding tokens, as well as our **Start-of-Sentence (SOS)** and **End-of-Sentence (EOS)** tokens. We keep a running count of the number of words in our vocabulary, too (which is 3 to start with as our corpus already contains the three tokens mentioned). These are the default values for an empty vocabulary; however, they will be populated as we read our data in:

```
PAD_token = 0
SOS_token = 1
EOS_token = 2

class Vocabulary:
    def __init__(self, name):
        self.name = name
        self.trimmed = False
```

```
        self.word2index = {}
        self.word2count = {}
        self.index2word = {PAD_token: "PAD", SOS_token:
                           "SOS", EOS_token: "EOS"}
        self.num_words = 3
```

2. Next, we create the functions that we will use to populate our vocabulary. addWord
    takes a word as input. If this is a new word that is not already in our vocabulary, we
    add this word to our indices, set the count of this word to 1, and increment the total
    number of words in our vocabulary by 1. If the word in question is already in our
    vocabulary, we simply increment the count of this word by 1:

```
    def addWord(self, w):
        if w not in self.word2index:
            self.word2index[w] = self.num_words
            self.word2count[w] = 1
            self.index2word[self.num_words] = w
            self.num_words += 1
        else:
            self.word2count[w] += 1
```

3. We also use the addSentence function to apply the addWord function to all the
    words within a given sentence:

```
    def addSentence(self, sent):
        for word in sent.split(' '):
            self.addWord(word)
```

One thing we can do to speed up the training of our model is reduce the size of
our vocabulary. This means that any embedding layers will be much smaller and
the total number of learned parameters within our model can be fewer. An easy
way to do this is to remove any low-frequency words from our vocabulary. Any
words occurring just once or twice in our dataset are unlikely to have huge
predictive power, and so removing them from our corpus and replacing them
with blank tokens in our final model could reduce the time taken for our model
to train and reduce overfitting without having much of a negative impact on our
model's predictions.

4. To remove low-frequency words from our vocabulary, we can implement a `trim` function. The function first loops through the word count dictionary and if the occurrence of the word is greater than the minimum required count, it is appended to a new list:

```
def trim(self, min_cnt):
    if self.trimmed:
        return
    self.trimmed = True

    words_to_keep = []

    for k, v in self.word2count.items():
        if v >= min_cnt:
            words_to_keep.append(k)

    print('Words to Keep: {} / {} = {:.2%}'.format(
        len(words_to_keep), len(self.word2index),
        len(words_to_keep) / len(self.word2index)))
```

5. Finally, our indices are rebuilt from the new `words_to_keep` list. We set all the indices to their initial empty values and then repopulate them by looping through our kept words with the `addWord` function:

```
self.word2index = {}
self.word2count = {}
self.index2word = {PAD_token: "PAD",\
                    SOS_token: "SOS",\
                    EOS_token: "EOS"}
self.num_words = 3

for w in words_to_keep:
    self.addWord(w)
```

We have now defined a vocabulary class that can be easily populated with our input sentences. Next, we actually need to load in our dataset to create our training data.

Loading the data

We will start loading in the data using the following steps:

1. The first step for reading in our data is to perform any necessary steps to clean the data and make it more human-readable. We start by converting it from Unicode into ASCII format. We can easily use a function to do this:

```
def unicodeToAscii(s):
    return ''.join(
        c for c in unicodedata.normalize('NFD', s)
        if unicodedata.category(c) != 'Mn'
    )
```

2. Next, we want to process our input strings so that they are all in lowercase and do not contain any trailing whitespace or punctuation, except the most basic characters. We can do this by using a series of regular expressions:

```
def cleanString(s):
    s = unicodeToAscii(s.lower().strip())
    s = re.sub(r"([.!?])", r" \1", s)
    s = re.sub(r"[^a-zA-Z.!?]+", r" ", s)
    s = re.sub(r"\s+", r" ", s).strip()
    return s
```

3. Finally, we apply this function within a wider function—readVocs. This function reads our data file into lines and then applies the cleanString function to every line. It also creates an instance of the Vocabulary class that we created earlier, meaning this function outputs both our data and vocabulary:

```
def readVocs(datafile, corpus_name):
    lines = open(datafile, encoding='utf-8').\
        read().strip().split('\n')
    pairs = [[cleanString(s) for s in l.split('\t')]
             for l in lines]
    voc = Vocabulary(corpus_name)
    return voc, pairs
```

Next, we filter our input pairs by their maximum length. This is again done to reduce the potential dimensionality of our model. Predicting sentences that are hundreds of words long would require a very deep architecture. In the interest of training time, we want to limit our training data here to instances where the input and output are less than 10 words long.

4. To do this, we create a couple of filter functions. The first one, `filterPair`, returns a Boolean value based on whether the current line has an input and output length that is less than the maximum length. Our second function, `filterPairs`, simply applies this condition to all the pairs within our dataset, only keeping the ones that meet this condition:

```
def filterPair(p, max_length):
    return len(p[0].split(' ')) < max_length and
len(p[1].split(' ')) < max_length


def filterPairs(pairs, max_length):
    return [pair for pair in pairs if filterPair(pair,
        max_length)]
```

5. Now, we just need to create one final function that applies all the previous functions we have put together and run it to create our vocabulary and data pairs:

```
def loadData(corpus, corpus_name, datafile, save_dir,
max_length):
    voc, pairs = readVocs(datafile, corpus_name)
    print(str(len(pairs)) + " Sentence pairs")
    pairs = filterPairs(pairs,max_length)
    print(str(len(pairs))+ " Sentence pairs after
        trimming")
    for p in pairs:
        voc.addSentence(p[0])
        voc.addSentence(p[1])
    print(str(voc.num_words) + " Distinct words in
        vocabulary")
    return voc, pairs


max_length = 10
voc, pairs = loadData(corpus, corpus_name, datafile,
                max_length)
```

We can see that our input dataset consists of over 200,000 pairs. When we filter this to sentences where both the input and output are less than 10 words long, this reduces to just 64,000 pairs consisting of 18,000 distinct words:

```
221282 Sentence pairs
64271 Sentence pairs after trimming
18008 Distinct words in vocabulary
```

Figure 8.7 – Value of sentences in the dataset

6. We can print a selection of our processed input/output pairs in order to verify that our functions have all worked correctly:

```
print("Example Pairs:")
for pair in pairs[-10:]:
    print(pair)
```

The following output is generated:

```
Example Pairs:
['four', 'three minutes to go !']
['three minutes to go !', 'yes .']
['another fifteen seconds to go .', 'do something ! stall them !']
['yes sir name please ?', 'food !']
['food !', 'do you have a reservation ?']
['do you have a reservation ?', 'food ! !']
['grrrhmmnnnjkjmmmnn !', 'franz ! help ! lunatic !']
['what o clock is it mr noggs ?', 'eleven o clock my lorj']
['stuart ?', 'yes .']
['yes .', 'how quickly can you move your artillery forward ?']
```

Figure 8.8 – Processed input/output pairs

It appears that we have successfully split our dataset into input and output pairs upon which we can train our network.

Finally, before we begin building the model, we must remove the rare words from our corpus and data pairs.

Removing rare words

As previously mentioned, including words that only occur a few times within our dataset will increase the dimensionality of our model, increasing our model's complexity and the time it will take to train the model. Therefore, it is preferred to remove them from our training data to keep our model as streamlined and efficient as possible.

You may recall earlier that we built a `trim` function into our vocabulary, which will allow us to remove infrequently occurring words from our vocabulary. We can now create a function to remove these rare words and call the `trim` method from our vocabulary as our first step. You will see that this removes a large percentage of words from our vocabulary, indicating that the majority of the words in our vocabulary occur infrequently. This is expected as the distribution of words within any language model will follow a long-tail distribution. We will use the following steps to remove the words:

1. We first calculate the percentage of words that we will keep within our model:

```
def removeRareWords(voc, all_pairs, minimum):
    voc.trim(minimum)
```

This results in the following output:

```
Words to Keep: 7823 / 18005 = 43.45%
```

Figure 8.9 – Percentage of words to be kept

2. Within this same function, we loop through all the words in the input and output sentences. If for a given pair either the input or output sentence has a word that isn't in our new trimmed corpus, we drop this pair from our dataset. We print the output and see that even though we have dropped over half of our vocabulary, we only drop around 17% of our training pairs. This again reflects how our corpus of words is distributed over our individual training pairs:

```
pairs_to_keep = []

for p in all_pairs:
    keep = True

    for word in p[0].split(' '):
        if word not in voc.word2index:
            keep = False
            break
    for word in p[1].split(' '):
        if word not in voc.word2index:
            keep = False
            break

    if keep:
```

```
        pairs_to_keep.append(p)

print("Trimmed from {} pairs to {}, {:.2%} of total".\
        format(len(all_pairs), len(pairs_to_keep),
            len(pairs_to_keep)/ len(all_pairs)))
return pairs_to_keep


minimum_count = 3
pairs = removeRareWords(voc, pairs, minimum_count)
```

This results in the following output:

```
Trimmed from 64271 pairs to 53165, 82.72% of total
```

Figure 8.10 – Final value after building our dataset

Now that we have our finalized dataset, we need to build some functions that transform our dataset into batches of tensors that we can pass to our model.

Transforming sentence pairs to tensors

We know that our model will not take raw text as input, but rather, tensor representations of sentences. We will also not process our sentences one by one, but instead in smaller batches. For this, we require both our input and output sentences to be transformed into tensors, where the width of the tensor represents the size of the batch that we wish to train on:

1. We start by creating several helper functions, which we can use to transform our pairs into tensors. We first create a indexFromSentence function, which grabs the index of each word in the sentence from the vocabulary and appends an EOS token to the end:

```
def indexFromSentence(voc, sentence):
    return [voc.word2index[word] for word in\
        sent.split(' ')] + [EOS_token]
```

2. Secondly, we create a zeroPad function, which pads any tensors with zeroes so that all of the sentences within the tensor are effectively the same length:

```
def zeroPad(l, fillvalue=PAD_token):
    return list(itertools.zip_longest(*l,\
        fillvalue=fillvalue))
```

3. Then, to generate our input tensor, we apply both of these functions. First, we get the indices of our input sentence, then apply padding, and then transform the output into `LongTensor`. We will also obtain the lengths of each of our input sentences out output this as a tensor:

```
def inputVar(l, voc):
    indexes_batch = [indexFromSentence(voc, sentence)\
                    for sentence in l]
    padList = zeroPad(indexes_batch)
    padTensor = torch.LongTensor(padList)
    lengths = torch.tensor([len(indexes) for indexes\
                            in indexes_batch])
    return padTensor, lengths
```

4. Within our network, our padded tokens should generally be ignored. We don't want to train our model on these padded tokens, so we create a Boolean mask to ignore these tokens. To do so, we use a `getMask` function, which we apply to our output tensor. This simply returns 1 if the output consists of a word and 0 if it consists of a padding token:

```
def getMask(l, value=PAD_token):
    m = []
    for i, seq in enumerate(l):
        m.append([])
        for token in seq:
            if token == PAD_token:
                m[i].append(0)
            else:
                m[i].append(1)
    return m
```

5. We then apply this to our `outputVar` function. This is identical to the `inputVar` function, except that along with the indexed output tensor and the tensor of lengths, we also return the Boolean mask of our output tensor. This Boolean mask just returns `True` when there is a word within the output tensor and `False` when there is a padding token. We also return the maximum length of sentences within our output tensor:

```python
def outputVar(l, voc):
    indexes_batch = [indexFromSentence(voc, sentence)
                     for sentence in l]
    max_target_len = max([len(indexes) for indexes in
                          indexes_batch])
    padList = zeroPad(indexes_batch)
    mask = torch.BoolTensor(getMask(padList))
    padTensor = torch.LongTensor(padList)
    return padTensor, mask, max_target_len
```

6. Finally, in order to create our input and output batches concurrently, we loop through the pairs in our batch and create input and output tensors for both pairs using the functions we created previously. We then return all the necessary variables:

```python
def batch2Train(voc, batch):
    batch.sort(key=lambda x: len(x[0].split(" ")),\
               reverse=True)

    input_batch = []
    output_batch = []

    for p in batch:
        input_batch.append(p[0])
        output_batch.append(p[1])

    inp, lengths = inputVar(input_batch, voc)
    output, mask, max_target_len = outputVar(output_
                                             batch, voc)

    return inp, lengths, output, mask, max_target_len
```

7. This function should be all we need to transform our training pairs into tensors for training our model. We can validate that this is working correctly by performing a single iteration of our `batch2Train` function on a random selection of our data. We set our batch size to 5 and run this once:

```
test_batch_size = 5
batches = batch2Train(voc, [random.choice(pairs) for _\
                                in range(test_batch_size)])
input_variable, lengths, target_variable, mask, max_
target_len = batches
```

Here, we can validate that our input tensor has been created correctly. Note how the sentences end with padding (0 tokens) where the sentence length is less than the maximum length for the tensor (in this instance, 9). The width of the tensor also corresponds to the batch size (in this case, 5):

```
Input:
tensor([[ 147,  147,  691, 1359, 1324],
        [ 582, 1125,    3, 1007,    4],
        [  94,   12,   98, 2444,    2],
        [   7,   54,   12,  173,    0],
        [  60,   84,  514,    4,    0],
        [ 441,    6,    4,    2,    0],
        [1114,    2,    2,    0,    0],
        [   6,    0,    0,    0,    0],
        [   2,    0,    0,    0,    0]])
```

Figure 8.11 – Input tensor

We can also validate the corresponding output data and mask. Notice how the `False` values in the mask overlap with the padding tokens (the zeroes) in our output tensor:

```
Target:
tensor([[  36,    27,   318,   690,    50],
        [1153,  1095,   572,    25,    37],
        [   4,     6,     4,   614,  1324],
        [   2,     2,     2,    53,   534],
        [   0,     0,     0,  1214,    40],
        [   0,     0,     0,     6,    47],
        [   0,     0,     0,     2,   169],
        [   0,     0,     0,     0,    76],
        [   0,     0,     0,     0,     6],
        [   0,     0,     0,     0,     2]])
Mask:
tensor([[ True,   True,   True,   True,   True],
        [ True,   True,   True,   True,   True],
        [ True,   True,   True,   True,   True],
        [ True,   True,   True,   True,   True],
        [False,  False,  False,   True,   True],
        [False,  False,  False,   True,   True],
        [False,  False,  False,   True,   True],
        [False,  False,  False,  False,   True],
        [False,  False,  False,  False,   True],
        [False,  False,  False,  False,   True]])
```

Figure 8.12 – The target and mask tensors

Now that we have obtained, cleaned, and transformed our data, we are ready to begin training the attention-based model that will form the basis of our chatbot.

Constructing the model

We start, as with our other sequence-to-sequence models, by creating our encoder. This will transform the initial tensor representation of our input sentence into hidden states.

Constructing the encoder

We will now create the encoder by taking the following steps:

1. As with all of our PyTorch models, we start by creating an Encoder class that inherits from nn.Module. All the elements here should look familiar to the ones used in previous chapters:

```
class EncoderRNN(nn.Module):
    def __init__(self, hidden_size, embedding,\
                 n_layers=1, dropout=0):
        super(EncoderRNN, self).__init__()
        self.n_layers = n_layers
        self.hidden_size = hidden_size
        self.embedding = embedding
```

Next, we create our **Recurrent Neural Network (RNN)** module. In this chatbot, we will be using a **Gated Recurrent Unit (GRU)** instead of the **Long Short-Term Memory (LSTM)** models we saw before. GRUs are slightly less complex than LSTMs as although they still control the flow of information through the RNN, they don't have separate forget and update gates like the LSTM. We use GRUs in this instance for a few main reasons:

a) GRUs have proven to be more computationally efficient as there are fewer parameters to learn. This means that our model will train much more quickly with GRUs than with LSTMs.

b) GRUs have proven to have similar performance levels over short sequences of data as LSTMs. LSTMs are more useful when learning longer sequences of data. In this instance we are only using input sentences with 10 words or less, so GRUs should produce similar results.

c) GRUs have proven to be more effective at learning from small datasets than LSTMs. As the size of our training data is small relative to the complexity of the task we're trying to learn, we should opt to use GRUs.

2. We now define our GRU, taking into account the size of our input, the number of layers, and whether we should implement dropout:

```
self.gru = nn.GRU(hidden_size, hidden_size, n_layers,
                dropout=(0 if n_layers == 1 else\
                        dropout), bidirectional=True)
```

Notice here how we implement bidirectionality into our model. You will recall from previous chapters that a bidirectional RNN allows us to learn from a sentence moving sequentially forward through a sentence, as well as moving sequentially backward. This allows us to better capture the context of each word in the sentence relative to those that appear before and after it. Bidirectionality in our GRU means our encoder looks something like this:

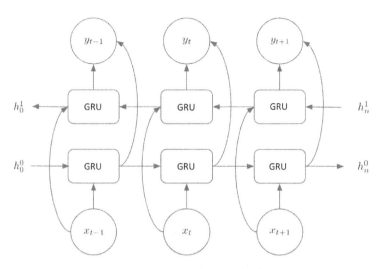

Figure 8.13 – Encoder layout

We maintain two hidden states, as well as outputs at each step, within our input sentence.

3. Next, we need to create a forward pass for our encoder. We do this by first embedding our input sentences and then using the `pack_padded_sequence` function on our embeddings. This function "packs" our padded sequence so that all of our inputs are of the same length. We then pass out the packed sequences through our GRU to perform a forward pass:

```
def forward(self, input_seq, input_lengths, hidden=None):
    embedded = self.embedding(input_seq)
    packed = nn.utils.rnn.pack_padded_sequence(embedded,
                                    input_lengths)
    outputs, hidden = self.gru(packed, hidden)
```

4. After this, we unpack our padding and sum the GRU outputs. We can then return this summed output, along with our final hidden state, to complete our forward pass:

```
outputs, _ = nn.utils.rnn.pad_packed_sequence(outputs)
outputs = outputs[:, :, :self.hidden_size] + a \
        outputs[:, : ,self.hidden_size:]
return outputs, hidden
```

Now, we will move on to creating an attention module in the next section.

Constructing the attention module

Next, we need to build our attention module, which we will apply to our encoder so that we can learn from the relevant parts of the encoder's output. We will do so as follows:

1. Start by creating a class for the attention model:

    ```
    class Attn(nn.Module):
        def __init__(self, hidden_size):
            super(Attn, self).__init__()
            self.hidden_size = hidden_size
    ```

2. Then, create the `dot_score` function within this class. This function simply calculates the dot product of our encoder output with the output of our hidden state by our encoder. While there are other ways of transforming these two tensors into a single representation, using a dot product is one of the simplest:

    ```
    def dot_score(self, hidden, encoder_output):
        return torch.sum(hidden * encoder_output, dim=2)
    ```

3. We then use this function within our forward pass. First, calculate the attention weights/energies based on the `dot_score` method, then transpose the results, and return the softmax transformed probability scores:

    ```
    def forward(self, hidden, encoder_outputs):
        attn_energies = self.dot_score(hidden, \
                                    encoder_outputs)
        attn_energies = attn_energies.t()
        return F.softmax(attn_energies, dim=1).unsqueeze(1)
    ```

Next, we can use this attention module within our decoder to create an attention-focused decoder.

Constructing the decoder

We will now construct the decoder, as follows:

1. We begin by creating our `DecoderRNN` class, inheriting from `nn.Module` and defining the initialization parameters:

    ```
    class DecoderRNN(nn.Module):
        def __init__(self, embedding, hidden_size, \
                    output_size, n_layers=1, dropout=0.1):
    ```

```
super(DecoderRNN, self).__init__()

self.hidden_size = hidden_size
self.output_size = output_size
self.n_layers = n_layers
self.dropout = dropout
```

2. We then create our layers within this module. We will create an embedding layer and a corresponding dropout layer. We use GRUs again for our decoder; however, this time, we do not need to make our GRU layer bidirectional as we will be decoding the output from our encoder sequentially. We will also create two linear layers—one regular layer for calculating our output and one layer that can be used for concatenation. This layer is twice the width of the regular hidden layer as it will be used on two concatenated vectors, each with a length of `hidden_size`. We also initialize an instance of our attention module from the last section in order to be able to use it within our `Decoder` class:

```
self.embedding = embedding
self.embedding_dropout = nn.Dropout(dropout)
self.gru = nn.GRU(hidden_size, hidden_size, n_layers,
    dropout=(0 if n_layers == 1 else dropout))
self.concat = nn.Linear(2 * hidden_size, hidden_size)
self.out = nn.Linear(hidden_size, output_size)

self.attn = Attn(hidden_size)
```

3. After defining all of our layers, we need to create a forward pass for the decoder. Notice how the forward pass will be used one step (word) at a time. We start by getting the embedding of the current input word and making a forward pass through the GRU layer to get our output and hidden states:

```
def forward(self, input_step, last_hidden, encoder_
outputs):
    embedded = self.embedding(input_step)
    embedded = self.embedding_dropout(embedded)
    rnn_output, hidden = self.gru(embedded, last_hidden)
```

4. Next, we use the attention module to get the attention weights from the GRU output. These weights are then multiplied by the encoder outputs to effectively give us a weighted sum of our attention weights and our encoder output:

```
attn_weights = self.attn(rnn_output, encoder_outputs)
context = attn_weights.bmm(encoder_outputs.transpose(0,
                                                        1))
```

5. We then concatenate our weighted context vector with the output of our GRU and apply a tanh function to get out final concatenated output:

```
rnn_output = rnn_output.squeeze(0)
context = context.squeeze(1)
concat_input = torch.cat((rnn_output, context), 1)
concat_output = torch.tanh(self.concat(concat_input))
```

6. For the final step within our decoder, we simply use this final concatenated output to predict the next word and apply a softmax function. The forward pass finally returns this output, along with the final hidden state. This forward pass will be iterated upon, with the next forward pass using the next word in the sentence and this new hidden state:

```
output = self.out(concat_output)
output = F.softmax(output, dim=1)
return output, hidden
```

Now that we have defined our models, we are ready to define the training process

Defining the training process

The first step of the training process is to define the measure of loss for our models. As our input tensors may consist of padded sequences, owing to our input sentences all being of different lengths, we cannot simply calculate the difference between the true output and the predicted output tensors. To account for this, we will define a loss function that applies a Boolean mask over our outputs and only calculates the loss of the non-padded tokens:

1. In the following function, we can see that we calculate cross-entropy loss across the whole output tensors. However, to get the total loss, we only average over the elements of the tensor that are selected by the Boolean mask:

```
def NLLMaskLoss(inp, target, mask):
    TotalN = mask.sum()
```

```
    CELoss = -torch.log(torch.gather(inp, 1,\
                        target.view(-1, 1)).squeeze(1))
    loss = CELoss.masked_select(mask).mean()
    loss = loss.to(device)
    return loss, TotalN.item()
```

2. For the majority of our training, we need two main functions—one function, `train()`, which performs training on a single batch of our training data and another function, `trainIters()`, which iterates through our whole dataset and calls `train()` on each of the individual batches. We start by defining `train()` in order to train on a single batch of data. Create the `train()` function, then get the gradients to 0, define the device options, and initialize the variables:

```
def train(input_variable, lengths, target_variable,\
          mask, max_target_len, encoder, decoder,\
          embedding, encoder_optimizer,\
          decoder_optimizer, batch_size, clip,\
          max_length=max_length):

    encoder_optimizer.zero_grad()
    decoder_optimizer.zero_grad()

    input_variable = input_variable.to(device)
    lengths = lengths.to(device)
    target_variable = target_variable.to(device)
    mask = mask.to(device)

    loss = 0
    print_losses = []
    n_totals = 0
```

3. Then, perform a forward pass of the inputs and sequence lengths though the encoder to get the output and hidden states:

```
encoder_outputs, encoder_hidden = encoder(input_variable,
lengths)
```

4. Next, we create our initial decoder input, starting with SOS tokens for each sentence. We then set the initial hidden state of our decoder to be equal to that of the encoder:

```
decoder_input = torch.LongTensor([[SOS_token for _ in \
                                    range(batch_size)]])
decoder_input = decoder_input.to(device)
decoder_hidden = encoder_hidden[:decoder.n_layers]
```

Next, we implement teacher forcing. If you recall from the last chapter, teacher forcing, when generating output sequences with some given probability, we use the true previous output token rather than the predicted previous output token to generate the next word in our output sequence. Using teacher forcing helps our model converge much more quickly; however, we must be careful not to make the teacher forcing ratio too high or else our model will be too reliant on the teacher forcing and will not learn to generate the correct output independently.

5. Determine whether we should use teacher forcing for the current step:

```
use_TF = True if random.random() < teacher_forcing_ratio
else False
```

6. Then, if we do need to implement teacher forcing, run the following code. We pass each of our sequence batches through the decoder to obtain our output. We then set the next input as the true output (target). Finally, we calculate and accumulate the loss using our loss function and print this to the console:

```
for t in range(max_target_len):
decoder_output, decoder_hidden = decoder(
  decoder_input, decoder_hidden, encoder_outputs)

decoder_input = target_variable[t].view(1, -1)
mask_loss, nTotal = NLLMaskLoss(decoder_output, \
      target_variable[t], mask[t])
loss += mask_loss
print_losses.append(mask_loss.item() * nTotal)
n_totals += nTotal
```

7. If we do not implement teacher forcing on a given batch, the procedure is almost identical. However, instead of using the true output as the next input into the sequence, we use the one generated by the model:

```
_, topi = decoder_output.topk(1)
decoder_input = torch.LongTensor([[topi[i][0] for i in \
                                 range(batch_size)]])
decoder_input = decoder_input.to(device)
```

8. Finally, as with all of our models, the final steps are to perform backpropagation, implement gradient clipping, and step through both of our encoder and decoder optimizers to update the weights using gradient descent. Remember that we clip out gradients in order to prevent the vanishing/exploding gradient problem, which was discussed in earlier chapters. Finally, our training step returns our average loss:

```
loss.backward()

_ = nn.utils.clip_grad_norm_(encoder.parameters(), clip)
_ = nn.utils.clip_grad_norm_(decoder.parameters(), clip)

encoder_optimizer.step()
decoder_optimizer.step()

return sum(print_losses) / n_totals
```

9. Next, as previously stated, we need to create the `trainIters()` function, which repeatedly calls our training function on different batches of input data. We start by splitting our data into batches using the `batch2Train` function we created earlier:

```
def trainIters(model_name, voc, pairs, encoder, decoder,\
               encoder_optimizer, decoder_optimizer,\
               embedding, encoder_n_layers, \
               decoder_n_layers, save_dir, n_iteration,\
               batch_size, print_every, save_every, \
               clip, corpus_name, loadFilename):
```

```
training_batches = [batch2Train(voc,\
                    [random.choice(pairs) for _ in\
                    range(batch_size)]) for _ in\
                    range(n_iteration)]
```

10. We then create a few variables that will allow us to count iterations and keep track of the total loss over each epoch:

```
print('Starting ...')
start_iteration = 1
print_loss = 0
if loadFilename:
    start_iteration = checkpoint['iteration'] + 1
```

11. Next, we define our training loop. For each iteration, we get a training batch from our list of batches. We then extract the relevant fields from our batch and run a single training iteration using these parameters. Finally, we add the loss from this batch to our overall loss:

```
print("Beginning Training...")
for iteration in range(start_iteration, n_iteration + 1):
    training_batch = training_batches[iteration - 1]
    input_variable, lengths, target_variable, mask, \
        max_target_len = training_batch

    loss = train(input_variable, lengths,\
                target_variable, mask, max_target_len,\
                encoder, decoder, embedding, \
                encoder_optimizer, decoder_optimizer,\
                batch_size, clip)
    print_loss += loss
```

12. On every iteration, we also make sure we print our progress so far, keeping track of how many iterations we have completed and what our loss was for each epoch:

```
if iteration % print_every == 0:
    print_loss_avg = print_loss / print_every
    print("Iteration: {}; Percent done: {:.1f}%;\
    Mean loss: {:.4f}".format(iteration,
                        iteration / n_iteration \
                        * 100, print_loss_avg))
    print_loss = 0
```

13. For the sake of completion, we also need to save our model state after every few epochs. This allows us to revisit any historical models we have trained; for example, if our model were to begin overfitting, we could revert back to an earlier iteration:

```
if (iteration % save_every == 0):
    directory = os.path.join(save_dir, model_name,\
                        corpus_name, '{}-{}_{}'.\
                        format(encoder_n_layers,\
                        decoder_n_layers, \
                        hidden_size))
    if not os.path.exists(directory):
        os.makedirs(directory)
    torch.save({
        'iteration': iteration,
        'en': encoder.state_dict(),
        'de': decoder.state_dict(),
        'en_opt': encoder_optimizer.state_dict(),
        'de_opt': decoder_optimizer.state_dict(),
        'loss': loss,
        'voc_dict': voc.__dict__,
        'embedding': embedding.state_dict()
        }, os.path.join(directory, '{}_{}.tar'.
format(iteration, 'checkpoint')))
```

Now that we have completed all the necessary steps to begin training our model, we need to create functions to allow us to evaluate the performance of the model.

Defining the evaluating process

Evaluating a chatbot is slightly different from evaluating other sequence-to-sequence models. In our text translation task, an English sentence will have one direct translation into German. While there may be multiple correct translations, for the most part, there is a single correct translation from one language into another.

For chatbots, there are multiple different valid outputs. Take the following three lines from some conversations with a chatbot:

Input: *"Hello"*

Output: *"Hello"*

Input: *"Hello"*

Output: *"Hello. How are you?"*

Input: *"Hello"*

Output: *"What do you want?"*

Here, we have three different responses, each one equally valid as a response. Therefore, at each stage of our conversation with our chatbot, there will be no single "correct" response. So, evaluation is much more difficult. The most intuitive way of testing whether a chatbot produces a valid output is by having a conversation with it! This means we need to set up our chatbot in a way that enables us to have a conversation with it to determine whether it is working well:

1. We will start by defining a class that will allow us to decode the encoded input and produce text. We do this by using what is known as a **greedy encoder**. This simply means that at each step of the decoder, our model takes the word with the highest predicted probability as the output. We start by initializing the GreedyEncoder() class with our pretrained encoder and decoder:

```
class GreedySearchDecoder(nn.Module):
    def __init__(self, encoder, decoder):
        super(GreedySearchDecoder, self).__init__()
        self.encoder = encoder
        self.decoder = decoder
```

2. Next, define a forward pass for our decoder. We pass the input through our encoder to get our encoder's output and hidden state. We take the encoder's final hidden layer to be the first hidden input to the decoder:

```
def forward(self, input_seq, input_length, max_length):
    encoder_outputs, encoder_hidden = \
                    self.encoder(input_seq, input_length)
    decoder_hidden = encoder_hidden[:decoder.n_layers]
```

3. Then, create the decoder input with SOS tokens and initialize the tensors to append decoded words to (initialized as a single zero value):

```
decoder_input = torch.ones(1, 1, device=device,
dtype=torch.long) * SOS_token
all_tokens = torch.zeros([0], device=device,
dtype=torch.long)
all_scores = torch.zeros([0], device=device)
```

4. After that, iterate through the sequence, decoding one word at a time. We perform a forward pass through the encoder and add a max function to obtain the highest-scoring predicted word and its score, which we then append to the all_tokens and all_scores variables. Finally, we take this predicted token and use it as the next input to our decoder. After the whole sequence has been iterated over, we return the complete predicted sentence:

```
for _ in range(max_length):
    decoder_output, decoder_hidden = self.decoder\
        (decoder_input, decoder_hidden, encoder_outputs)
    decoder_scores, decoder_input = \
        torch.max (decoder_output, dim=1)
    all_tokens = torch.cat((all_tokens, decoder_input),\
                    dim=0)
    all_scores = torch.cat((all_scores, decoder_scores),\
                    dim=0)
    decoder_input = torch.unsqueeze(decoder_input, 0)
return all_tokens, all_scores
```

All the pieces are beginning to come together. We have the defined training and evaluation functions, so the final step is to write a function that will actually take our input as text, pass it to our model, and obtain a response from the model. This will be the "interface" of our chatbot, where we actually have our conversation.

5. We first define an `evaluate()` function, which takes our input function and returns the predicted output words. We start by transforming our input sentence into indices using our vocabulary. We then obtain a tensor of the lengths of each of these sentences and transpose it:

```
def evaluate(encoder, decoder, searcher, voc, sentence,\
             max_length=max_length):
    indices = [indexFromSentence(voc, sentence)]
    lengths = torch.tensor([len(indexes) for indexes \
                            in indices])
    input_batch = torch.LongTensor(indices).transpose(0, 1)
```

6. Then, we assign our lengths and input tensors to the relevant devices. Next, run the inputs through the searcher (`GreedySearchDecoder`) to obtain the word indices of the predicted output. Finally, we transform these word indices back into word tokens before returning them as the function output:

```
input_batch = input_batch.to(device)
lengths = lengths.to(device)
tokens, scores = searcher(input_batch, lengths, \
                          max_length)
decoded_words = [voc.index2word[token.item()] for \
                 token in tokens]
return decoded_words
```

7. Finally, we create a `runchatbot` function, which acts as the interface with our chatbot. This function takes human-typed input and prints the chatbot's response. We create this function as a `while` loop that continues until we terminate the function or type `quit` as our input:

```
def runchatbot(encoder, decoder, searcher, voc):
    input_sentence = ''
    while(1):
        try:
            input_sentence = input('> ')
            if input_sentence == 'quit': break
```

8. We then take the typed input and normalize it, before passing the normalized input to our `evaluate()` function, which returns the predicted words from the chatbot:

```
input_sentence = cleanString(input_sentence)
output_words = evaluate(encoder, decoder, searcher,\
                  voc, input_sentence)
```

9. Finally, we take these output words and format them, ignoring the EOS and padding tokens, before printing the chatbot's response. Because this is a `while` loop, this allows us to continue the conversation with the chatbot indefinitely:

```
output_words[:] = [x for x in output_words if \
                 not (x == 'EOS' or x == 'PAD')]
print('Response:', ' '.join(output_words))
```

Now that we have constructed all the functions necessary to train, evaluate, and use our chatbot, it's time to begin the final step—training our model and conversing with our trained chatbot.

Training the model

As we have defined all the necessary functions, training the model just becomes a case or initializing our hyperparameters and calling our training functions:

1. We first initialize our hyperparameters. While these are only suggested hyperparameters, our models have been set up in a way that will allow them to adapt to whatever hyperparameters they are passed. It is good practice to experiment with different hyperparameters to see which ones result in an optimal model configuration. Here, you could experiment with increasing the number of layers in your encoder and decoder, increasing or decreasing the size of the hidden layers, or increasing the batch size. All of these hyperparameters will have an effect on how well your model learns, as well as a number of other factors, such as the time it takes to train the model:

```
model_name = 'chatbot_model'
hidden_size = 500
encoder_n_layers = 2
decoder_n_layers = 2
dropout = 0.15
batch_size = 64
```

2. After that, we can load our checkpoints. If we have previously trained a model, we can load the checkpoints and model states from previous iterations. This saves us from having to retrain our model each time:

```
loadFilename = None
checkpoint_iter = 4000


if loadFilename:
    checkpoint = torch.load(loadFilename)
    encoder_sd = checkpoint['en']
    decoder_sd = checkpoint['de']
    encoder_optimizer_sd = checkpoint['en_opt']
    decoder_optimizer_sd = checkpoint['de_opt']
    embedding_sd = checkpoint['embedding']
    voc.__dict__ = checkpoint['voc_dict']
```

3. After that, we can begin to build our models. We first load our embeddings from the vocabulary. If we have already trained a model, we can load the trained embeddings layer:

```
embedding = nn.Embedding(voc.num_words, hidden_size)

if loadFilename:
    embedding.load_state_dict(embedding_sd)
```

4. We then do the same for our encoder and decoder, creating model instances using the defined hyperparameters. Again, if we have already trained a model, we simply load the trained model states into our models:

```
encoder = EncoderRNN(hidden_size, embedding, \
                    encoder_n_layers, dropout)
decoder = DecoderRNN(embedding, hidden_size, \
                    voc.num_words, decoder_n_layers,
                    dropout)

if loadFilename:
    encoder.load_state_dict(encoder_sd)
    decoder.load_state_dict(decoder_sd)
```

5. Last but not least, we specify a device for each of our models to be trained on. Remember, this is a crucial step if you wish to use GPU training:

```
encoder = encoder.to(device)
decoder = decoder.to(device)
print('Models built and ready to go!')
```

If this has all worked correctly and your models have been created with no errors, you should see the following:

```
Building encoder and decoder ...
Models built and ready to go!
```

Figure 8.14 – Successful output

Now that we have created instances of both our encoder and decoders, we are ready to begin training them.

We start by initializing some training hyperparameters. In the same way as our model hyperparameters, these can be adjusted to influence training time and how our model learns. Clip controls the gradient clipping and teacher forcing controls how often we use teacher forcing within our model. Notice how we use a teacher forcing ratio of 1 so that we always use teacher forcing. Lowering the teaching forcing ratio would mean our model takes much longer to converge; however, it might help our model generate correct sentences by itself better in the long run.

6. We also need to define the learning rates of our models and our decoder learning ratio. You will find that your model performs better when the decoder carries out larger parameter updates during gradient descent. Therefore, we introduce a decoder learning ratio to apply a multiplier to the learning rate so that the learning rate is greater for the decoder than it is for the encoder. We also define how often our model prints and saves the results, as well as how many epochs we want our model to run for:

```
save_dir = './'


clip = 50.0
teacher_forcing_ratio = 1.0
learning_rate = 0.0001
decoder_learning_ratio = 5.0


epochs = 4000


print_every = 1
save_every = 500
```

7. Next, as always when training models in PyTorch, we switch our models to training mode to allow the parameters to be updated:

```
encoder.train()
decoder.train()
```

8. Next, we create optimizers for both the encoder and decoder. We initialize these as Adam optimizers, but other optimizers will work equally well. Experimenting with different optimizers may yield different levels of model performance. If you have trained a model previously, you can also load the optimizer states if required:

```
print('Building optimizers ...')
encoder_optimizer = optim.Adam(encoder.parameters(), \
                            lr=learning_rate)
decoder_optimizer = optim.Adam(decoder.parameters(),
                lr=learning_rate * decoder_learning_ratio)


if loadFilename:
    encoder_optimizer.load_state_dict(\
                                encoder_optimizer_sd)
    decoder_optimizer.load_state_dict(\
                                decoder_optimizer_sd)
```

9. The final step before running the training is to make sure CUDA is configured to be called if you wish to use GPU training. To do this, we simply loop through the optimizer states for both the encoder and decoder and enable CUDA across all of the states:

```
for state in encoder_optimizer.state.values():
    for k, v in state.items():
        if isinstance(v, torch.Tensor):
            state[k] = v.cuda()


for state in decoder_optimizer.state.values():
    for k, v in state.items():
        if isinstance(v, torch.Tensor):
            state[k] = v.cuda()
```

10. Finally, we are ready to train our model. This can be done by simply calling the `trainIters` function with all the required parameters:

```
print("Starting Training!")
trainIters(model_name, voc, pairs, encoder, decoder,\
            encoder_optimizer, decoder_optimizer, \
            embedding, encoder_n_layers, \
            decoder_n_layers, save_dir, epochs, \
            batch_size,print_every, save_every, \
            clip, corpus_name, loadFilename)
```

If this is working correctly, you should see the following output start to print:

```
Building optimizers ...
Starting Training!
Initializing ...
Training...
Iteration: 1; Percent complete: 0.0%; Average loss: 8.3863
Iteration: 2; Percent complete: 0.1%; Average loss: 7.7756
Iteration: 3; Percent complete: 0.1%; Average loss: 7.1381
Iteration: 4; Percent complete: 0.1%; Average loss: 6.8835
Iteration: 5; Percent complete: 0.1%; Average loss: 6.7475
```

Figure 8.15 – Training the model

Your model is now training! Depending on a number of factors, such as how many epochs you have set your model to train for and whether you are using a GPU, your model may take some time to train. When it is complete, you will see the following output. If everything has worked correctly, your model's average loss will be significantly lower than when you started training, showing that your model has learned something useful:

```
Iteration: 3996; Percent complete: 99.9%; Average loss: 2.5443
Iteration: 3997; Percent complete: 99.9%; Average loss: 2.6106
Iteration: 3998; Percent complete: 100.0%; Average loss: 2.7075
Iteration: 3999; Percent complete: 100.0%; Average loss: 2.8199
Iteration: 4000; Percent complete: 100.0%; Average loss: 2.8155
```

Figure 8.16 – Average loss after 4,000 iterations

Now that our model has been trained, we can begin the evaluation process and start using our chatbot.

Evaluating the model

Now that we have successfully created and trained our model, it is time to evaluate its performance. We will do so by taking the following steps:

1. To begin the evaluation, we first switch our model into evaluation mode. As with all other PyTorch models, this is done to prevent any further parameter updates occurring within the evaluation process:

```
encoder.eval()
decoder.eval()
```

2. We also initialize an instance of GreedySearchDecoder in order to be able to perform the evaluation and return the predicted output as text:

```
searcher = GreedySearchDecoder(encoder, decoder)
```

3. Finally, to run the chatbot, we simply call the runchatbot function, passing it encoder, decoder, searcher, and voc:

```
runchatbot(encoder, decoder, searcher, voc)
```

Doing so will open up an input prompt for you to enter your text:

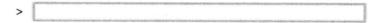

Figure 8.17 – UI element for entering text

Entering your text here and pressing *Enter* will send your input to the chatbot. Using our trained model, our chatbot will create a response and print it to the console:

```
> hello
Response: hello . . . . .
```

Figure 8.18 – Output for the chatbot

You can repeat this process as many times as you like to have a "conversation" with the chatbot. At a simple conversational level, the chatbot can produce surprisingly good results:

```
> hello
Response: hello . . . . .
> how are you?
Response: fine . i m fine . you ?
> I'm good
Response: good . you re a good man .
```

Figure 8.19 – Output for the chatbot

However, once the conversation gets more complex, it will become obvious that the chatbot isn't capable of the same level of conversation as a human:

```
> who are you?
Response: i m in the middle of the way .
> in the middle of what?
Response: i m not talking about her .
```

Figure 8.20 – Limitations of the chatbot

In many cases, your chatbot's responses may be nonsensical:

```
> what's one plus one?
Response: they re not gonna hurt me . .
```

Figure 8.21 – Wrong output

It is clear that we have created a chatbot capable of simple back and forth conversations. However, we still have a long way to go before our chatbot is able to pass the Turing test and be able to convince us that we are actually talking to a human being. However, considering the relatively small corpus of data we have trained our model on, the use of attention in our sequence-to-sequence model has shown reasonably good results, demonstrating just how versatile these architectures can be.

While the best chatbots are trained on vast corpuses of billions of data points, our model has proven reasonably effective with a relatively small one. However, basic attention networks are no longer state-of-the-art and in our next chapter, we will discuss some of the more recent developments for NLP learning that have resulted in even more realistic chatbots.

Summary

In this chapter, we applied all the knowledge we learned from our recurrent models and our sequence-to-sequence models and combined them with an attention mechanism to construct a fully working chatbot. While conversing with our chatbot is unlikely to be indistinguishable from talking to a real human, with a considerably larger dataset we might hope to achieve an even more realistic chatbot.

Although sequence-to-sequence models with attention were state-of-the-art in 2017, machine learning is a rapidly progressing field and since then, there have been multiple improvements made to these models. In the final chapter, we will discuss some of these state-of-the-art models in more detail, as well as cover several other contemporary techniques used in machine learning for NLP, many of which are still in development.

9
The Road Ahead

The field of machine learning is rapidly expanding, with new revelations being made almost yearly. The field of machine learning for NLP is no exception, with advancements being made rapidly and the performance of machine learning models on NLP tasks incrementally increasing.

So far in this book, we have discussed a number of machine learning methodologies that allow us to build models to perform NLP tasks such as classification, translation, and approximating conversation via a chatbot. However, as we have seen so far, the performance of our models has been worse and relative to that of a human being. Even using the techniques we have examined so far, including sequence-to-sequence networks with attention mechanisms, we are unlikely to train a chatbot model that will match or outperform a real person. However, we will see in this chapter that recent developments in the field of NLP have been made that bring us one step closer to the goal of creating chatbots that are indistinguishable from humans.

In this chapter, we will explore a couple of state-of-the art machine learning models for NLP and examine some of the features that result in superior performance. We will then turn to look at several other NLP tasks that are currently the focus of much research, and how machine learning techniques might be used to solve them.

In this chapter, we will cover the following topics:

- Exploring state-of-the-art NLP machine learning
- Future NLP tasks
- Semantic role labeling
- Constituency parsing
- Textual entailment
- Machine comprehension

Exploring state-of-the-art NLP machine learning

While the techniques we have learned in this book so far are highly useful methodologies for training our own machine learning model from scratch, they are far from the most sophisticated models being developed globally. Companies and research groups are constantly striving to create the most advanced machine learning models that will achieve the highest performance on a number of NLP tasks.

Currently, there are two NLP models that have the best performance and could be considered state-of-the-art: **BERT** and **GPT-2**. Both models are forms of **generalized language models**. We will discuss these in more detail in the upcoming sections.

BERT

BERT, which stands for **Bidirectional Encoder Representations from Transformers**, was developed by Google in 2018 and is widely considered to be the leading model in the field of NLP, having achieved leading performance in natural language inference and question-answering tasks. Fortunately, this has been released as an open source model, so this can be downloaded and used for NLP tasks of your own.

BERT was released as a pre-trained model, which means users can download and implement BERT without the need to retrain the model from scratch each time. The pre-trained model is trained on several corpuses, including the whole of Wikipedia (consisting of 2.5 billion words) and another corpus of books (which includes a further 800 million words). However, the main element of BERT that sets it apart from other similar models is the fact that it provides a deep, bidirectional, unsupervised language representation, which is shown to provide a more sophisticated, detailed representation, thus leading to improved performance in NLP tasks.

Embeddings

While traditional embedding layers (such as GLoVe) form a single representation of a word that is agnostic to the meaning of the word within the sentence, the bidirectional BERT model attempts to form representations based on its context. For example, in these two sentences, the word *bat* has two different meanings.

"The bat flew past my window"

"He hit the baseball with the bat"

Although the word *bat* is a noun in both sentences, we can discern that the context and meaning of the word is obviously very different, depending on the other words around it. Some words may also have different meanings, depending on whether they are a noun or verb within the sentence:

"She used to match to light the fire"

"His poor performance meant they had no choice but to fire him"

Using the bidirectional language model to form context-dependent representations of words is what truly makes BERT stand out as a state-of-the-art model. For any given token, we obtain its input representation by combining the token, position, and segment embeddings:

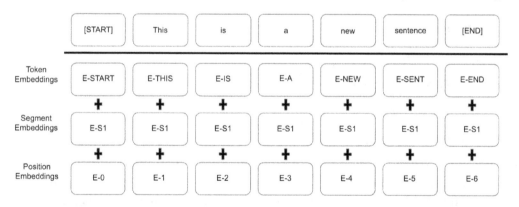

Figure 9.1 – BERT architecture

However, it is important to understand how the model arrives at these initial context-dependent token-embeddings.

Masked language modeling

In order to create this bidirectional language representation, BERT uses two different techniques, the first of which is masked language modeling. This methodology effectively hides 15% of the words within the input sentences by replacing them with a masking token. The model then tries to predict the true values of the masked words, based on the context of the other words in the sentence. This prediction is made bidirectionally in order to capture the context of the sentence in both directions:

Input: *We [MASK_1] hide some of the [MASK_2] in the sentence*

Labels: *MASK_1 = randomly, MASK_2 = words*

If our model can learn to predict the correct context-dependent words, then we are one step closer to context-dependent representation.

Next sentence prediction

The other technique that BERT uses to learn the language representation is next sentence prediction. In this methodology, our model receives two sentences and our model learns to predict whether the second sentence is the sentence that follows the first sentence; for example:

Sentence A: *"I like to drink coffee"*

Sentence B: *"It is my favorite drink"*

Is Next Sentence?: *True*

Sentence A: *"I like to drink coffee"*

Sentence B: *"The sky is blue"*

Is Next Sentence?: *False*

By passing our model pairs of sentences like this, it can learn to determine whether any two sentences are related and follow one another, or whether they are just two random, unrelated sentences. Learning these sentence relationships is useful in a language model as many NLP-related tasks, such as question-answering, require the model to understand the relationship between two sentences. Training a model on next sentence prediction allows the model to identify some kind of relationship between a pair of sentences, even if that relationship is very basic.

BERT is trained using both methodologies (masked language modeling and next sentence prediction), and the combined loss function of both techniques is minimized. By using two different training methods, our language representation is sufficiently robust and learns how sentences are formed and structured, as well as how different sentences relate to one another.

BERT–Architecture

The model architecture builds upon many of the principles we have seen in the previous chapters to provide a sophisticated language representation using bidirectional encoding. There are two different variants of BERT, each consisting of a different number of layers and attention heads:

- **BERT Base**: 12 transformer blocks (layers), 12 attention heads, ~110 million parameters
- **BERT Large**: 24 transformer blocks (layers), 16 attention heads, ~340 million parameters

While BERT Large is just a deeper version of BERT Base with more parameters, we will focus on the architecture of BERT Base.

BERT is built by following the principle of a **transformer**, which will now be explained in more detail.

Transformers

The model architecture builds upon many of the principles we have seen so far in this book. By now, you should be familiar with the concept of encoders and decoders, where our model learns an encoder to form a representation of an input sentence, and then learns a decoder to decode this representation into a final output, whether this be a classification or translation task:

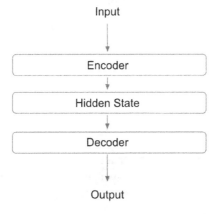

Figure 9.2 – Transformer workflow

However, our transformer adds another element of sophistication to this approach, where a transformer actually has a stack of encoders and a stack of decoders, with each decoder receiving the output of the final encoder as its input:

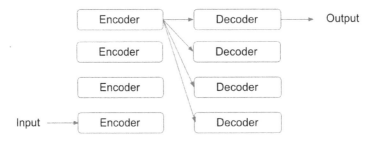

Figure 9.3 – Transformer workflow for multiple encoders

Within each encoder layer, we find two constituent parts: a self-attention layer and a feed-forward layer. The self-attention layer is the layer that receives the model's input first. This layer causes the encoder to examine other words within the input sentence as it encodes any received word, making the encoding context aware. The output from the self-attention layer is passed forward to a feed-forward layer, which is applied independently to each position. This can be illustrated diagrammatically like so:

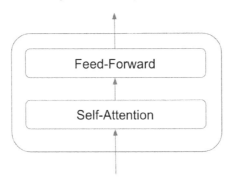

Figure 9.4 – Feedforward layer

Our decoder layers are almost identical in structure to our encoders, but they incorporate an additional attention layer. This attention layer helps the decoder focus on the relevant part of the encoded representation, similar to how we saw attention working within our sequence-to-sequence models:

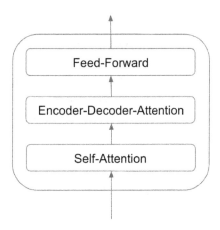

Figure 9.5 – Attention methodology

We know that our decoders take input from our final encoder, so one linked encoder/ decoder might look something like this:

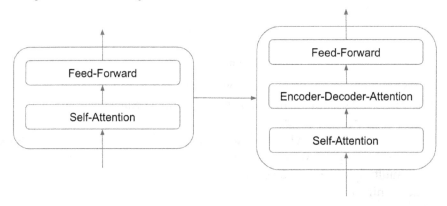

Figure 9.6 – Linked encoder/decoder array

This should provide you with a useful overview of how the different encoders and decoders are stacked up within the larger model. Next, we will examine the individual parts in more detail.

Encoders

The unique property of transformers is that words flow through the encoder layers individually and each word in each position has its own path. While there are some dependencies within the self-attention layer, these don't exist within the feed-forward layer. The vectors for the individual words are obtained from an embedding layer and then fed through a self-attention layer before being fed through a feed-forward network:

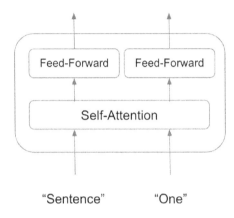

Figure 9.7 – Encoder layout

Self-attention is arguably the most complex component of the encoder, so we will examine this in more detail first. Let's say we have a three-word input sentence; for example, *"This is fine"*. For each word within this sentence, we represent them as a single word vector that was obtained from the embedding layer of our model. We then extract three vectors from this single word vector: a query vector, a key vector, and a value vector. These three vectors are obtained by multiplying our word vector by three different weight matrices that are obtained while training the model.

If we call our word embeddings for each word in our input sentence, *Ethis*, *Eis*, and *Efine*, we can calculate our query, key, and value vectors like so:

Query vectors:

$$W^q = Learned\ Query\ Weight\ Matrix$$

$$Q_{this} = W^q * E_{this}$$

$$Q_{is} = W^q * E_{is}$$

$$Q_{fine} = W^q * E_{fine}$$

Key vectors:

$$W^k = Learned\ Key\ Weight\ Matrix$$

$$k_{this} = W^k * E_{this}$$

$$k_{is} = W^k * E_{is}$$

$$k = W^k * E_{fine}$$

Value vectors:

$$W^v = Learned\ Value\ Weight\ Matrix$$

$$v_{this} = W^v * E_{this}$$

$$v_{is} = W^v * E_{is}$$

$$v_{fine} = W^v * E_{fine}$$

Now that we know how to calculate each of these vectors, it is important to understand what each of them represents. Essentially, each of these is an abstraction of a concept within the attention mechanism. This will become apparent once we see how they are calculated.

Let's continue with our working example. We need to consider each word within our input sentence in turn. To do this, we calculate a score for each pair of query/key vectors in our sentence. This is done by obtaining the dot product of each query/key vector pair for each word within our input sentence. For example, to calculate the scores for the first word in the sentence, "this", we calculate the dot product between the query vector for "this" and the key vector in position 0. We repeat this for the key vectors in all other positions within the input sentence, so we obtain n scores for the first word in our input sentence, where n is the length of the sentence:

Scores ("this"):

$$S^0{}_{this} = q_{this} \cdot k_{this}$$

$$S^1{}_{this} = q_{this} \cdot k_{is}$$

$$S^2{}_{this} = q_{this} \cdot k_{fine}$$

Next, we apply a softmax function to each of these scores so that the value of each is now between 0 and 1 (as this helps prevent exploding gradients and makes gradient descent more efficient and easily calculable). We then multiply each of these scores by the value vectors and sum these all up to obtain a final vector, which is then passed forward within the encoder:

Final vector ("this"):

$$V^0 = S^0_{this} \cdot V_{this}$$

$$V^1 = S^1_{this} * V_{is}$$

$$V^2 = S^2_{this} * V_{fine}$$

$$Final = V^0 + V^1 + V^2$$

We then repeat this procedure for all the words within the input sentence so that we obtain a final vector for each word, incorporating an element of self-attention, which is then passed along the encoder to the feed-forward network. This self-attention process means that our encoder knows where to look within the input sentence to obtain the information it needs for the task.

In this example, we only learned a single matrix of weights for our query, key, and value vectors. However, we can actually learn multiple different matrices for each of these elements and apply these simultaneously across our input sentence to obtain our final outputs. This is what's known as **multi-headed attention** and allows us to perform more complex attention calculations, relying on multiple different learned patterns rather than just a single attention mechanism.

We know that BERT incorporates 12 attention heads, meaning that 12 different weight matrices are learned for Wq, Wk, and Wv.

Finally, we need a way for our encoders to account for the order of words in the input sequence. Currently, our model treats each word in our input sequence independently, but in reality, the order of the words in the input sequence will make a huge difference to the overall meaning of the sentence. To account for this, we use **positional encoding**.

To apply this, our model takes each input embedding and adds a positional encoding vector to each one individually. These positional vectors are learned by our model, following a specific pattern to help them determine the position of each word in the sequence. In theory, adding these positional vectors to our initial embeddings should translate into meaningful distances between our final vectors, once they are projected into the individual query, key, and value vectors:

x0 = Raw Embedding

t0 = Positional Encoding

E0 = Embedding with Time Signal

x0 + t0 = E0

Our model learns different positional encoding vectors for each position (t_0, t_1, and so on), which we then apply to each word in our input sentence before these even enter our encoder:

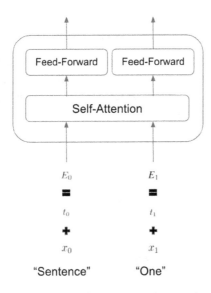

Figure 9.8 – Adding input to the encoder

Now that we have covered the main components of the encoder, it's time to look at the other side of the model and see how the decoder is constructed.

Decoders

The components in decoders are much the same of those in encoders. However, rather than receiving the raw input sentence like encoders do, the decoders in our transformer receive their inputs from the outputs of our encoders.

Our stacked encoders process our input sentence and we are left with a set of attention vectors, *K* and *V*, which are used within the encoder-decoder attention layer of our decoder. This allows it to focus only on the relevant parts of the input sequence:

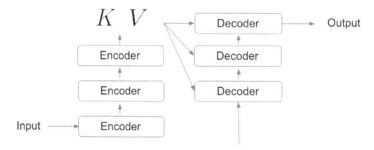

Figure 9.9 – Stacked decoders

At each time step, our decoders use a combination of the previously generated words in the sentence and the *K,V* attention vectors to generate the next word in the sentence. This process is repeated iteratively until the decoder generates an <END> token, indicating that it has completed generating the final output. One given time step on the transformer decoder may look like this:

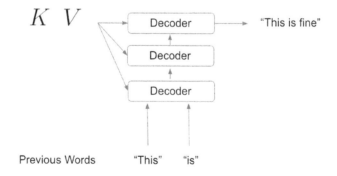

Figure 9.10 – Transformer decoder

It is worth noting here that the self-attention layers within the decoders operate in a slightly different way to those found in our encoders. Within our decoder, the self-attention layer only focuses on earlier positions within the output sequence. This is done by masking any future positions of the sequence by setting them to minus infinity. This means that when the classification happens, the softmax calculation always results in a prediction of 0.

The encoder-decoder attention layer works in the same way as the multi-headed self-attention layer within our encoder. However, the main difference is that it creates a query matrix from the layer below and takes the key and values matrix from the output of the encoders.

These encoder and decoder parts comprise our transformer, which forms the basis for BERT. Next, we will look at some of the applications of BERT and examine a few variations that have shown increased performance at specific tasks.

Applications of BERT

Being state-of-the-art, BERT obviously has a number of practical applications. Currently, it is being used in a number of Google products that you probably use on a daily basis; namely, suggested replies and smart compose in Gmail (where Gmail predicts your expected sentence based on what you are currently typing) and autocomplete within the Google search engine (where you type the first few characters you wish to search for and the drop-down list will predict what you are going to search for).

As we saw in the previous chapter, chatbots are one of the most impressive things NLP deep learning can be used for, and the use of BERT has led to some very impressive chatbots indeed. In fact, question-answering is one of the main things that BERT excels at, largely due to the fact that it is trained on a large knowledge base (Wikipedia) and is able to answer questions in a syntactically correct way (due to being trained with next sentence prediction in mind).

We are still not at the stage where chatbots are indistinguishable from conversations with real humans, and the ability of BERT to draw from its knowledge base is extremely limited. However, some of the results achieved by BERT are promising and, taking into account how quickly the field of NLP machine learning is progressing, this suggests that this may become a reality very soon.

Currently, BERT is only able to address a very narrow type of NLP task due to the way it is trained. However, there are many variations of BERT that have been changed in subtle ways to achieve increased performance at specific tasks. These include, but are not limited to, the following:

- **roBERTa**: A variation of BERT, built by Facebook. Removes the next sentence prediction element of BERT, but enhances the word masking strategy by implementing dynamic masking.

- **xlm/BERT**: Also built by Facebook, this model applies a dual-language training mechanism to BERT that allows it to learn relationships between words in different languages. This allows BERT to be used effectively for machine translation tasks, showing improved performance over basic sequence-to-sequence models.

- **distilBERT**: A more compact version of BERT, retaining 95% of the original but halving the number of learned parameters, reducing the model's total size and training time.

- **ALBERT**: This Google trained model uses its own unique training method called sentence order prediction. This variation of BERT has been shown to outperform the standard BERT across a number of tasks and is now considered state-of-the-art ahead of BERT (illustrating just how quickly things can change!).

While BERT is perhaps the most well known, there are also other transformer-based models that are considered state-of-the-art. The major one that is often considered a rival to BERT is GPT-2.

GPT-2

GPT-2, while similar to BERT, differs in some subtle ways. While both models are based upon the transformer architecture previously outlined, BERT uses a form of attention known as self-attention, while GPT-2 uses masked self-attention. Another subtle difference between the two is that GPT-2 is constructed in such a way that it can to output one token at a time.

This is because GPT-2 is essentially auto-regressive in the way it works. This means that when it generates an output (the first word in a sentence), this output is added recursively to the input. This input is then used to predict the next word in the sentence and is repeated until a complete sentence has been generated. You can see this in the following example:

Step 1:

Input: *What color is the sky?*

Output: *The ...*

We then add the predicted output to the end of our input and repeat this step:

Step 2:

Input: *What color is the sky? The*

Output: *sky*

We repeat this process until we have generated the entire sentence:

Step 3:

Input: *What color is the sky? The sky*

Output: *is*

Step 4:

Input: *What color is the sky? The sky is*

Output: *blue*

This is one of the key trade-offs in terms of performance between BERT and GPT-2. The fact that BERT is trained bidirectionally means this single-token generation is not possible; however, GPT-2 is not bidirectional, so it only considers previous words in the sentence when making predictions, which is why BERT outperforms GPT-2 when predicting missing words within a sentence.

Comparing self-attention and masked self-attention

This difference is also apparent in the way the two different models implement attention. Since BERT is bidirectional, its attention mechanism is able to consider the context of the whole input sentence and determine where exactly in the input sentence to look:

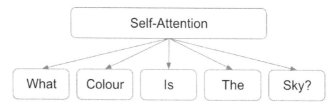

Figure 9.11 – Self-Attention mechanism

On the other hand, GPT-2 is not bidirectional, so the masked self-attention mechanism is only able to look at tokens it has already seen and is not able to "look ahead":

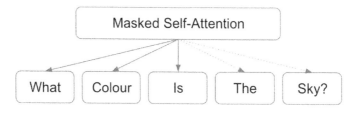

Figure 9.12 – Masked self-attention mechanism

GPT-2 – Ethics

One very interesting consideration of GPT-2 was the author's consideration of ethics when considering whether to release the model to the public or not.

The language generation capabilities of GPT-2 mean that it is capable of generating very realistic and syntactically correct sentences using very few prompts. By providing it with just a couple of lines of input to identify a subject matter, the model is able to generate paragraphs of detailed information about this subject, regardless of whether it is actually true or not! One, now infamous, example was the "Unicorn" story, where the model received one sentence input:

Input: *"In a shocking finding, scientists discovered a herd of unicorns living in a remote, previously unexplored valley, in the Andes Mountains. Even more surprising to the researchers was the fact that the unicorns spoke perfect English."*

The model gave the following as output:

Output: *The scientist named the population, after their distinctive horn, Ovid's Unicorn. These four-horned, silver-white unicorns were previously unknown to science.*

Now, after almost two centuries, the mystery of what sparked this odd phenomenon is finally solved.

Dr. Jorge Pérez, an evolutionary biologist from the University of La Paz, and several companions, were exploring the Andes Mountains when they found a small valley, with no other animals or humans. Pérez noticed that the valley had what appeared to be a natural fountain, surrounded by two peaks of rock and silver snow.

This is only the first three paragraphs of a much larger nine-paragraph response that the model generated from very little input. What's notable is that the sentences all make perfect sense (regardless of the impossible subject matter!), that the paragraphs flow together in a logical order, and that the model was able to generate all of this from a very small input.

While this is extremely impressive in terms of performance and what it's possible to achieve from building deep NLP models, it does raise some concerns about the ethics of such models and how they can be used (and abused!).

With the rise of "fake news" and the spread of misinformation using the internet, examples like this illustrate how simple it is to generate realistic text using these models. Let's consider an example where an agent wishes to generate fake news on a number of subjects online. Now, they don't even need to write the fake information themselves. In theory, they could train NLP models to do this for them, before disseminating this fake information on the internet. The authors of GPT-2 paid particular attention to this when training and releasing the model to the public, noting that the model had the potential to be abused and misused, therefore only releasing the larger more sophisticated models to the public once they saw no evidence of misuse of the smaller models.

This may become a key focus of NLP deep learning moving forward. As we approach chatbots and text generators such as GPT-2 that can approach human levels of sophistication, the uses and misuses of these models need to be fully understood. Studies have shown that GPT-2 generated text was deemed to be almost as credible (72%) as real human-written articles from the New York Times (83%). As we continue to develop even more sophisticated deep NLP models in the future, these numbers are likely to converge as model-generated text becomes more and more realistic moving forward.

Furthermore, the authors of GPT-2 also demonstrated that the model can be fine-tuned for misuse. By fine-tuning GPT-2 on ideologically extreme positions and generating text, it was shown that propaganda text can be generated, which supports these ideologies. While it was also shown that counter-models could be trained to detect these model-generated texts, we may again face further problems here in the future as these models become even more sophisticated.

These ethical considerations are worth keeping in mind as NLP models become even more complex and performant over time. While the models you train for your own purposes may not have been intended for any misuse, there is always the possibility that they could be used for purposes that were unintended. Always consider the potential applications of any model that you use.

Future NLP tasks

While the majority of this book has been focused on text classification and sequence generation, there are a number of other NLP tasks that we haven't really touched on. While many of these are more interesting from an academic perspective rather than a practical perspective, it's important to understand these tasks as they form the basis of how language is constructed and formed. Anything we, as NLP data scientists, can do to better understand the formation of natural language will only improve our understanding of the subject matter. In this section, we will discuss, in more detail, four key areas of future development in NLP:

- Constituency parsing
- Semantic role labeling
- Textual entailment
- Machine comprehension

Constituency parsing

Constituency parsing (also known as syntactic parsing) is the act of identifying parts of a sentence and assigning a syntactic structure to it. This syntactic structure is largely determined by the use of context-free grammars, meaning that using syntactic parsing, we can identify the underlying grammatical structure of a given sentence and map it out. Any sentence can be broken down into a "parse tree," which is a graphical representation of this underlying sentence structure, while syntactic parsing is the methodology by which this underlying structure is detected and determines how this tree is built.

We will begin by discussing this underlying grammatical structure. The idea of a "constituency" within a sentence is somewhat of an abstraction, but the basic assumption is that a sentence consists of multiple "groups" of words, each one of which is a constituency. Grammar, in its basic form, can be said to be an index of all possible types of constituencies that can occur within a sentence.

Let's first consider the most basic type of constituent, the **noun phrase**. Nouns within a sentence are fairly simple to identify as they are words that define objects or entities. In the following sentence, we can identify three nouns:

"Jeff the chef cooks dinner"

Jeff - Proper noun, denotes a name

Chef - A chef is an entity

Dinner - Dinner is an object/thing

However, a noun phrase is slightly different as each noun phrase should refer to one single entity. In the preceding sentence, even though *Jeff* and *chef* are both nouns, the phrase *Jeff the chef* refers to one single person, so this can be considered a noun phrase. But how can we determine syntactically that the noun phrase refers to a single entity? One simple way is to place the phrase before a verb and see if the sentence makes syntactic sense. If it does, then chances are, the phrase is a noun phrase:

Jeff the chef cooks...

Jeff the chef runs...

Jeff the chef drinks...

There exist a variety of different phrases that we are able to identify, as well as a number of complex grammatical rules that help us to identify them. We first identify the individual grammatical features that each sentence can be broken down into:

Grammar	Description	Examples
Noun	An object, thing, or entity	Boat, Cat, Soldier
Verb	A doing word	Jump, Run, Fight
Adjective	A descriptive word	Red, Sharp, Hot
Pronoun	Substitutes for a noun phrase	I, You, Him
Proper noun	A name	John, Italy, Mercedes
Determiner	Affixes to a noun to add context	The, This, These
Preposition	Describes temporal or spatial relationships	On, Under, To
Conjunction	Used to conjugate elements of the sentence together	And, But, Or

Now that we know that sentences are composed of constituents, and that constituents can be made up of several individual grammars, we can now start to map out our sentences based on their structure. For example, take the following example sentence:

"The boy hit the ball"

We can start by breaking this sentence down into two parts: a noun phrase and a verb phrase:

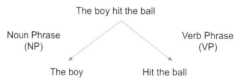

Figure 9.13 – Breaking down a sentence into its grammatical components

We then repeat this process for each of the phrases to split them into even smaller grammatical components. We can split this noun phrase into a determiner and a noun:

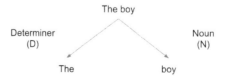

Figure 9.14 – Breaking down the noun phrase

Again, we do this for the verb phrase to break it down into a verb and another noun phrase:

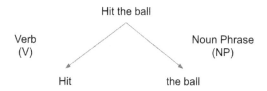

Figure 9.15 – Breaking down the verb phrase

We can iterate again and again, breaking down the various parts of our sentence into smaller and smaller chunks until we are left with a **parse tree**. This parse tree conveys the entirety of the syntactic structure of our sentence. We can see the parse tree of our example in its entirety here:

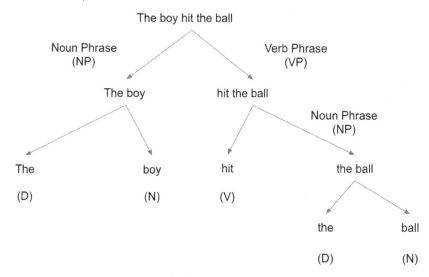

Figure 9.16 – Parse tree of the sentence

While these parse trees allow us to see the syntactic structure of our sentences, they are far from perfect. From this structure, we can clearly see that there are two noun phrases with a verb taking place. However, from the preceding structure, it is not clear what is actually taking place. We have an action between two objects, but it is not clear from syntax alone what is taking place. Which party is doing the action to whom? We will see that some of this ambiguity is captured by semantic role labeling.

Semantic role labeling

Semantic role labeling is the process of assigning labels to words or phrases within a sentence that indicates their semantic role within a sentence. In broad terms, this involves identifying the predicate of the sentence and determining how each of the other terms within the sentence are related to this predicate. In other words, for a given sentence, semantic role labeling determines "Who did what to whom and where/when?"

So, for a given sentence, we can generally break down a sentence into its constituent parts, like so:

Figure 9.17 Breaking down a sentence into its constituent parts

These parts of a sentence have specific semantic roles. The **predicate** of any given sentence represents the event occurring within the sentence, while all the other parts of the sentence relate back to a given predicate. In this sentence, we can label our "Who" as the agent of the predicate. The **agent** is the thing that causes the event. We can also label our "Whom" as the theme of our predicate. The **theme** is the element of our sentence most affected by the event in question:

Figure 9.18 – Breaking down the roles

In theory, each word or phrase in a sentence can be labeled with its specific semantic component. An almost comprehensive table for this is as follows:

Semantic Role	Definition	Example
Agent	The volitional causer of an event	The boy kicked the ball.
Experiencer	The experience of an event	Tom has a headache.
Force	The non-volitional causer of an event	The snow makes us cold.
Theme	The thing most affected by the event	He broke the glass.
Result	The final product of an event	They built a house.
Content	The content of a propositional event	She asked "Why?".
Instrument	The instrument/device used within an event	He caught a fish using a rod.
Beneficiary	The beneficiary of the event	She did it for him.
Source	The origin of an object or event	I drove from London.
Goal	The destination of an object or event	I drove to London.

By performing semantic role labeling, we can assign a specific role to every part of a sentence. This is very useful in NLP as it allows a model to "understand" a sentence better so that rather than a sentence just being an assortment of roles, it is understood as a combination of semantic roles that better convey what is actually happening in the event being described by the sentence.

When we read the sentence *"The boy kicked the ball"*, we inherently know that there is a boy, there is a ball, and that the boy is kicking the ball. However, all the NLP models we have looked at so far would comprehend this sentence by looking at the individual words in the sentence and creating some representation of them. It is unlikely that the fact that there are two "things" and that object one (the boy) is performing some action (kicking) on object two (the ball) would be understood by the systems we have seen so far. Introducing an element of semantic roles to our models could better help our systems form more realistic representations of sentences by defining the subjects of the sentences and the interactions between them.

One thing that semantic role labeling helps with greatly is the identification of sentences that convey the same meaning but are not grammatically or syntactically the same; such as the following, for example:

The man bought the apple from the shop

The shop sold the man an apple

The apple was bought by the man from the shop

The apple was sold by the shop to the man

These sentences have essentially the same meaning, although they clearly do not contain all the same words in the same order. By applying semantic role labeling to these sentences, we can determine that the predicate/agent/theme are all the same.

We previously saw how constituency parsing/syntactic parsing can be used to identify the syntactic structure of a sentence. Here, we can see how we can break down the simple sentence "I bought a cat" into its constituent parts – pronoun, verb, determinant, and noun:

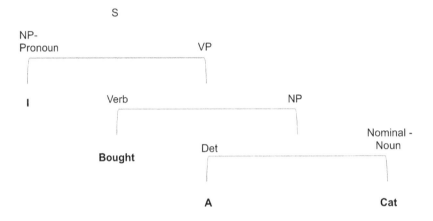

Figure 9.19 – Constituency parsing

However, this does not shed any insight on the semantic role each part of the sentence is playing. Is the cat being bought by me or am I being bought by the cat? While the syntactic role is useful for understanding the structure of the sentence, it doesn't shed as much light on the semantic meaning. A useful analogy is that of image captioning. In a model trained to label images, we would hope to achieve a caption that describes what is in an image. Semantic labeling is the opposite of this, where we take a sentence and try to abstract a mental "image" of what action is taking place in the sentence.

But what context is semantic role labeling useful for in NLP? In short, any NLP task that requires an element of "understanding" the content of text can be enhanced by the addition of roles. This could be anything from document summarization, question-answering, or sentence translation. For example, using semantic role labeling to identify the predicate of our sentence and the related semantic components, we could train a model to identify the components that contribute essential information to the sentence and drop those that do not.

Therefore, being able to train models to perform accurate and efficient semantic role labeling would have useful applications for the rest of NLP. The earliest semantic role labeling systems were purely rule-based, consisting of basic sets of rules derived from grammar. These have since evolved to incorporate statistical modeling approaches before the recent developments in deep learning, which meant that it is possible to train classifiers to identify the relevant semantic roles within a sentence.

As with any classification task, this is a supervised learning problem that requires a fully annotated sentence in order to train a model that will identify the semantic roles of previously unseen sentences. However, the availability of such annotated sentences is highly scarce. The gigantic language models, such as BERT, that we saw earlier in this chapter are trained on raw sentences and do not require labeled examples. However, in the case of semantic role labeling, our models require the use of correctly labeled sentences to be able to perform this task. While datasets do exist for this purpose, they are not large and versatile enough to train a fully comprehensive, accurate model that will perform well on a variety of sentences.

As you can probably imagine, the latest state-of-the-art approaches to solving the semantic role labeling task have all been neural network-based. Initial models used LSTMs and bidirectional LSTMs combined with GLoVe embeddings in order to perform classification on sentences. There have also been variations of these models that incorporate convolutional layers, which have also shown good performance.

However, it will be no surprise to learn that these state-of-the-art models are BERT-based. Using BERT has shown exemplary performance in a whole variety of NLP-related tasks, and semantic role labeling is no exception. Models incorporating BERT have been trained holistically to predict part-of-speech tags, perform syntactic parsing, and perform semantic role labeling simultaneously and have shown good results.

Other studies have also shown that graph convolutional networks are effective at semantic labeling. Graphs are constructed with nodes and edges, where the nodes within the graph represent semantic constituents and the edges represent the relationships between parent and child parts.

A number of open source models for semantic role labeling are also available. The SLING parser from Google is trained to perform semantic annotations of data. This model uses a bidirectional LSTM to encode sentences and a transition-based recurrent unit for decoding. The model simply takes text tokens as input and outputs roles without any further symbolic representation:

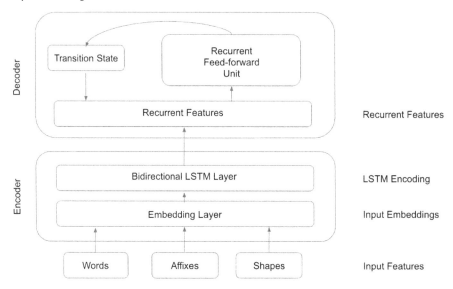

Figure 9.20 – Bi-directional LSTM (SLING)

It is worth noting that SLING is still a work in progress. Currently, it is not sophisticated enough to extract facts accurately from arbitrary texts. This indicates that there is much work to be done in the field before a true, accurate semantic role parser can be created. When this is done, a semantic role parser could easily be used as part of an ensemble machine learning model to label semantic roles within a sentence, which is then used within a wider machine learning model to enhance the model's "understanding" of text.

Textual entailment

Textual entailment is another methodology by which we can train models in an attempt to better understand the meaning of a sentence. In textual entailment, we attempt to identify a directional relationship between two pieces of text. This relationship exists whenever the truth from one piece of text follows from another piece of text. This means that, given two texts, if the second text can be held to be true by the information within the first text, we can say that there is a positive directional relationship between these two texts.

This task is often set up in the following fashion, with our first text labeled as text and our second text labeled as our hypothesis:

Text: *If you give money to charity, you will be happy*

Hypothesis: *Giving money to charity has good consequences*

This is an example of **positive textual entailment**. If the hypothesis follows from the text, then there can be said to be a directional relationship between the two texts. It is important to set up the example with a text/hypothesis as this defines the direction of the relationship. The majority of the time, this relationship is not symmetrical. For example, in this example, sentence one entails sentence two (we can infer sentence two to be true based on the information in sentence one). However, we cannot infer that sentence one is true based on the information in sentence two. While it is possible that both statements are indeed true, if we cannot deduce that there is a directional relationship between the two, we cannot infer one from the other.

There also exists a **negative textual entailment**. This is when the statements are contradictory; such as the following, for example:

Text: *If you give money to charity, you will be happy*

Hypothesis: *Giving money to charity has bad consequences*

In this example, the text does not entail the hypothesis; instead, the text contradicts the hypothesis. Finally, it is also possible to determine that there is **no textual entailment** between two sentences if there is no relationship between them. This means that the two statements are not necessarily contradictory, but rather that the text does not entail the hypothesis:

Text: *If you give money to charity, you will be happy*

Hypothesis: *Giving money to charity will make you relaxed*

The ambiguity of natural language makes this an interesting task from an NLP perspective. Two sentences can have a different syntactic structure, a different semantic structure, and consist of entirely different words but still have very similar meanings. Similarly, two sentences can consist of the same words and entities but have very different meanings.

This is where using models to be able to quantify the meaning of text is particularly useful. Textual entailment is also a unique problem in that two sentences may not have exactly the same meaning, yet one can still be inferred from the other. This requires an element of linguistic deduction that is not present in most language models. By incorporating elements of linguistic deduction in our models going forward, we can better capture the meaning of texts, as well as be able to determine whether two texts contain the same information, regardless of whether their representations are similar.

Fortunately, simple textual entailment models are not difficult to create, and LSTM-based models have been shown to be effective. One setup that may prove effective is that of a Siamese LSTM network.

We set up our model as a multi-class classification problem where two texts can be positively or negatively entailed or have no entailment. We feed our two texts into a dual-input model, thereby obtaining embeddings for the two texts, and pass them through bidirectional LSTM layers. The two outputs are then compared somehow (using some tensor operation) before they're fed through a final LSTM layer. Finally, we perform classification on the output using a softmax layer:

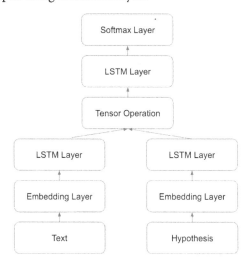

Figure 9.21 – Siamese LSTM network

While these models are far from perfect, they represent the first steps toward creating a fully accurate textual entailment model and open up the possibilities toward integrating this into language models moving forward.

Machine comprehension

So far in this book, we have referred mostly to NLP, but being able to process language is just part of the picture. When you or I read a sentence, we not only read, observe, and process the individual words, but we also build an inherent understanding of what the sentence actually means. Being able to train models that not only comprehend a sentence but can also form an understanding of the ideas being expressed within it is arguably the next step in NLP. The true definition of this field is very loosely defined, but it is often referred to as machine comprehension or **natural language understanding** (NLU).

At school, we are taught reading comprehension from a young age. You probably learned this skill a long time ago and is something you now take for granted. Often, you probably don't even realize you are doing it; in fact, you are doing it right now! Reading comprehension is simply the act of reading a text, understanding this text, and being able to answer questions about the text. For example, take a look at the following text:

> *As a method of disinfecting water, bringing it to its boiling point at 100 °C (212 °F) is the oldest and most effective way of doing this since it does not affect its taste. It is effective despite contaminants or particles present in it, and is a single step process that eliminates most microbes responsible for causing intestine-related diseases. The boiling point of water is 100 °C (212 °F) at sea level and at normal barometric pressure.*

Given that you understand this text, you should now be able to answer the following questions about it:

Q: *What is the boiling point of water?*

A: *100 °C (212 °F)*

Q: *Does boiling water affect its taste?*

A: *No*

This ability to understand text and answer questions about it form the basis for our machine comprehension task. We wish to be able to train a machine learning model that can not only form an understanding of a text, but also be able to answer questions about it in grammatically correct natural language.

The benefits of this are numerous, but a very intuitive use case would be to build a system that acts as a knowledge base. Currently, the way search engines work is that we run a search (in Google or a similar search engine) and the search engine returns a selection of documents. However, to find a particular piece of information, we must still infer the correct information from our returned document. The entire process might look something like this:

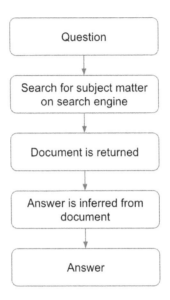

Figure 9.22 – Process of finding information

In this example, to answer the question *"What is the boiling point of water?"*, we first formulate our question. Then, we search for the subject matter on a search engine. This would probably be some reduced representation of the question; for example, *"water boiling point"*. Our search engine would then return some relevant documents, most likely the Wikipedia entry for water, which we would then manually have to search and use it to infer the answer to our question. While this methodology is effective, machine comprehension models would allow this process to be streamlined somewhat.

Let's say we have a perfect model that is able to fully comprehend and answer questions on a text corpus. We could train this model on a large source of data such as a large text scrape of the internet or Wikipedia and form a model that acts as a large knowledge base. By doing this, we would then be able to query the knowledge base with real questions and the answers would be returned automatically. This removes the knowledge inference step of our diagram as the inference is taken care of by the model as the model already has an understanding of the subject matter:

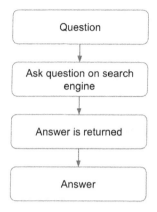

Figure 9.23 – New process using a model

In an ideal world, this would be as simple as typing *"What is the boiling point of water?"* into a search engine and receiving *"100 °C (212 °F)"* back as an answer.

Let's assume we have a simplified version of this model to begin with. Let's assume we already know the document that the answer to our asked question appears in. So, given the Wikipedia page on water, can we train a model to answer the question *"What is the boiling point of water?"*. A simple way of doing this to begin with, rather than incorporating the elements of a full language model, would be to simply return the passage of the Wikipedia page that contains the answer to our question.

An architecture that we could train to achieve this task might look something like this:

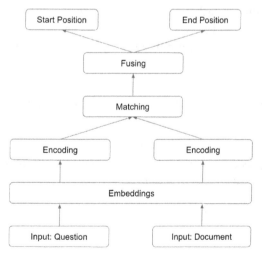

Figure 9.24 – Architecture of the model

Our model takes our question that we want answered and our document that contains our question as inputs. These are then passed through an embedding layer to form a tensor-based representation of each, and then an encoding layer to form a further reduced vector representation.

Now that our question and documents are represented as vectors, our matching layer attempts to determine where in the document vectors we should look to obtain the answer to our question. This is done through a form of attention mechanism whereby our question determines what parts of our document vectors we should look at in order to answer the question.

Finally, our fusing layer is designed to capture the long-term dependencies of our matching layer, combine all the received information from our matching layer, and perform a decoding step to obtain our final answers. This layer takes the form of a bidirectional RNN that decodes our matching layer output into final predictions. We predict two values here – a start point and an endpoint – using a multiclass classification. This represents the start and end points within our document that contain the answer to our initial question. If our document contained 100 words and the sentence between word 40 and word 50 contained the answer to our question, our model would ideally predict 40 and 50 for the values of the start and end points. These values could then be easily used to return the relevant passage from the input document.

While returning relevant areas of a target document is a useful model to train, it is not the same as a true machine comprehension model. In order to do that, we must incorporate elements of a larger language model.

In any machine comprehension task, there are actually three elements at play. We already know that there is a question and answer, but there is also a relevant context that may determine the answer to a given question. For example, we can ask the following question:

What day is it today?

The answer may differ, depending on the context in which the question is asked; for example, Monday, Tuesday, March the 6th, Christmas Day.

We must also note that the relationship between the question and answer is bidirectional. When given a knowledge base, it is possible for us to generate an answer given a question, but it also follows that we are able to generate a question given an answer:

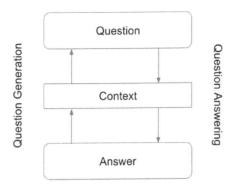

Figure 9.25 – Relationship between the question and answer

A true machine comprehension may be able to perform **question generation (QG)**, as well as **question-answering (QA)**. The most obvious solution to this is to train two separate models, one for each task, and compare their results. In theory, the output of our QG model should equal the input of our QA model, so by comparing the two, we can provide simultaneous evaluation:

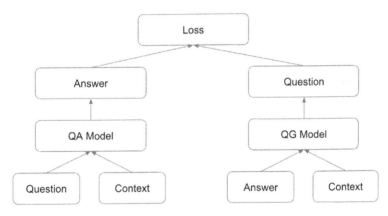

Figure 9.26 – Comparison between QG and QA models

However, a more comprehensive model would be able to perform these two tasks simultaneously, thereby generating a question from an answer and answering a question, much like humans are able to do:

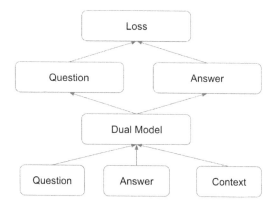

Figure 9.27 – Dual model representation

In fact, recent advances in NLU have meant that such models are now a reality. By combining many elements, we are able to create a neural network structure that is able to perform the function of the dual model, as illustrated previously. This is known as the **dual ask-answer network**. In fact, our model contains most of the components of neural networks that we have seen in this book so far, that is, embedding layers, convolutional layers, encoders, decoders, and attention layers. The full architecture of the ask-answer network looks similar to the following:

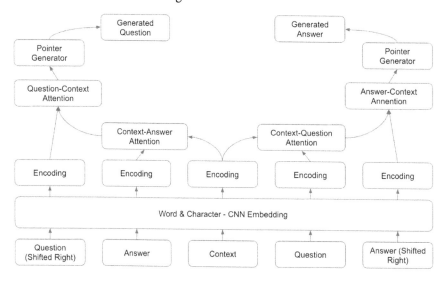

Figure 9.28 – Architecture of ask-answer network

We can make the following observations here:

- The model's **inputs** are the question, answer, and context, as previously outlined, but also the question and answer shifted right ward.

- Our **embedding** layer convolves across GLoVe embedded vectors for characters and words in order to create a combined representation.

- Our **encoders** consist of LSTMs, with applied attention.

- Our **outputs** are also RNN-based and decode our output one word at a time to generate final questions and answers.

While pre-trained ask-answer networks exist, you could practice implementing your newly acquired PyTorch skills and try building and training a model like this yourself.

Language comprehension models like these are likely to be one of the major focuses of study within NLP over the coming years, and new papers are likely to be published with great frequency moving forward.

Summary

In this chapter, we first examined several state-of-the-art NLP language models. BERT, in particular, seems to have been widely accepted as the industry standard state-of-the-art language model, and BERT and its variants are widely used by businesses in their own NLP applications.

Next, we examined several areas of focus for machine learning moving forward; namely semantic role labeling, constituency parsing, textual entailment, and machine comprehension. These areas will likely make up a large percentage of the current research being conducted in NLP moving forward.

Now that you have a well-rounded ability and understanding when it comes to NLP deep learning models and how to implement them in PyTorch, perhaps you'll feel inclined to be a part of this research moving forward. Whether this is in an academic or business context, you now hopefully know enough to create your own deep NLP projects from scratch and can use PyTorch to create the models you need to solve any NLP task you require. By continuing to improve your skills and by being aware and keeping up to date with all the latest developments in the field, you will surely be a successful, industry leading NLP data scientist!

Other Books You May Enjoy

If you enjoyed this book, you may be interested in these other books by Packt:

Hands-On Python Natural Language Processing

Aman Kedia, Mayank Rasu

ISBN: 978-1-83898-959-0

- Understand how NLP powers modern applications

- Explore key NLP techniques to build your natural language vocabulary

- Transform text data into mathematical data structures and learn how to improve text mining models

- Discover how various neural network architectures work with natural language data

- Get the hang of building sophisticated text processing models using machine learning and deep learning

- Check out state-of-the-art architectures that have revolutionized research in the NLP domain

Natural Language Processing with Python Quick Start Guide

Nirant Kasliwal

ISBN: 978-1-78913-038-6

- Understand classical linguistics in using English grammar for automatically generating questions and answers from a free text corpus

- Work with text embedding models for dense number representations of words, subwords and characters in the English language for exploring document clustering

- Deep Learning in NLP using PyTorch with a code-driven introduction to PyTorch

- Using an NLP project management Framework for estimating timelines and organizing your project into stages

- Hack and build a simple chatbot application in 30 minutes

- Deploy an NLP or machine learning application using Flask as RESTFUL APIs

Leave a review - let other readers know what you think

Please share your thoughts on this book with others by leaving a review on the site that you bought it from. If you purchased the book from Amazon, please leave us an honest review on this book's Amazon page. This is vital so that other potential readers can see and use your unbiased opinion to make purchasing decisions, we can understand what our customers think about our products, and our authors can see your feedback on the title that they have worked with Packt to create. It will only take a few minutes of your time, but is valuable to other potential customers, our authors, and Packt. Thank you!

Index

W

www.ingramcontent.com/pod-product-compliance
Lightning Source LLC
Chambersburg PA
CBHW080632060326
40690CB00021B/4901